KV-038-912

International Federation of Automatic Control

CONTROL APPLICATIONS OF NONLINEAR PROGRAMMING

Other Titles in the IFAC Proceedings Series

ATHERTON: Multivariable Technological Systems
BANKS & PRITCHARD: Control of Distributed Parameter Systems
CICHOCKI & STRASZAK: Systems Analysis Applications to Complex Programs
CRONHJORT: Real Time Programming 1978
CUENOD: Computer Aided Design of Control Systems
De GIORGO & ROVEDA: Criteria for Selecting Appropriate Technologies under Different Cultural, Technical and Social Conditions
DUBUISSON: Information and Systems
GHONAIMY: Systems Approach for Development
HARRISON: Distributed Computer Control Systems
HASEGAWA & INOUE: Urban, Regional and National Planning - Environmental Aspects
ISERMANN: Identification and System Parameter Estimation
LAUBER: Safety of Computer Control Systems
LEONHARD: Control in Power Electronics and Electrical Drives
MUNDAY: Automatic Control in Space
NIEMI: A Link Between Science and Applications of Automatic Control
NOVAK: Software for Computer Control
OSHIMA: Information Control Problems in Manufacturing Technology (1977)
REMBOLD: Information Control Problems in Manufacturing Technology (1979)
RIJNSDORP: Case Studies in Automation related to Humanization of Work
SAWARAGI & AKASHI: Environmental Systems Planning, Design and Control
SINGH & TITLI: Control and Management of Integrated Industrial Complexes
SINGH & TITLI: Large Scale Systems: Theory and Applications
SMEDEMA: Real Time Programming 1977
TOMOV: Optimization Methods - Applied Aspects

CONTROL APPLICATIONS OF NONLINEAR PROGRAMMING

Proceedings of the IFAC Workshop, Denver, Colorado, USA
21 June 1979

Edited by

H. E. RAUCH

Lockheed Palo Alto Research Laboratory, California, USA

Published for the

INTERNATIONAL FEDERATION OF AUTOMATIC CONTROL

by

PERGAMON PRESS

OXFORD · NEW YORK · TORONTO · SYDNEY · PARIS · FRANKFURT

U.K.	Pergamon Press Ltd., Headington Hill Hall, Oxford OX3 0BW, England
U.S.A.	Pergamon Press Inc., Maxwell House, Fairview Park, Elmsford, New York 10523, U.S.A.
CANADA	Pergamon of Canada, Suite 104, 150 Consumers Road, Willowdale, Ontario M2J 1P9, Canada
AUSTRALIA	Pergamon Press (Aust.) Pty. Ltd., P.O. Box 544, Potts Point, N.S.W. 2011, Australia
FRANCE	Pergamon Press SARL, 24 rue des Ecoles, 75240 Paris, Cedex 05, France
FEDERAL REPUBLIC OF GERMANY	Pergamon Press GmbH, 6242 Kronberg-Taunus, Hammerweg 6, Federal Republic of Germany

First edition 1980

British Library Cataloguing in Publication Data

IFAC Workshop on Control Applications of Nonlinear Programming, *Denver, 1979.*
Control applications of nonlinear programming.
1. Automatic control - Mathematical models - Congresses.
2. Nonlinear programming - Congresses
I. Title II. Rauch, H E
III. International Federation of Automatic Control.
629.8'312 TJ213 80-49944

ISBN 0-08-024491-2

These proceedings were reproduced by means of the photo-offset process using the manuscripts supplied by the authors of the different papers. The manuscripts have been typed using different typewriters and typefaces. The lay-out, figures and tables of some papers did not agree completely with the standard requirements; consequently the reproduction does not display complete uniformity. To ensure rapid publication this discrepancy could not be changed; nor could the English be checked completely. Therefore, the readers are asked to excuse any deficiencies of this publication which may be due to the above mentioned reasons.

The Editor

Printed in Great Britain by A. Wheaton & Co. Ltd, Exeter

CONTENTS

PREFACE

During the last decade there has been extensive theoretical development of numerical methods in nonlinear programming for parameter optimization and control. The increasing use of these methods has led to the organization of an international Working Group on Control Applications of Nonlinear Programming, under the auspices of the International Federation of Automatic Control. The purposes of the Working Group are to exchange information on the application of optimal and nonlinear programming techniques to real-life control problems, to investigate new ideas that arise from these exchanges, and to look for advances in optimal and nonlinear programming which are useful in solving modern control problems.

This volume contains the Proceedings of the first Workshop which was held in Denver, Colorado, U.S.A., on June 21, 1979. It represents the latest work of fifteen invited specialists from the U.S.A., U.S.S.R., Federal Republic of Germany, France, and Great Britain. The volume covers a variety of specific applications ranging from microprocessor control of automotive engines and optimal design of structures to optimal aircraft trajectories, system identification, and robotics. These significant contributions to numerical methods in control reflect the great amount of work done by the authors.

The impetus for the Workshop was derived from early efforts of Professor Henry Kelley (U.S.A.), Chairman of the IFAC Mathematics of Control Committee. We were fortunate to have had as members of the International Program Committee Professors Arthur Bryson, Jr. (U.S.A.), Faina M. Kirillova (U.S.S.R.), and R. W. H. Sargent (Great Britain). Finally, it was a great pleasure for me to have organized the Workshop and served as Chairman of the Working Group.

Herbert E. Rauch
Palo Alto, California U.S.A.

APPLICATION OF NON LINEAR PROGRAMMING TO OPTIMUM DESIGN PROBLEM

C. Knopf-Lenoir*, G. Touzot* and J. P. Yvon**

**Université de Technologie de Compiègne, BP 233, 60206 Compiègne, France*
***IRIA-LABORIA, BP 105, 78105 Le Chesnay, France*

Abstract. As a particular application of the optimal control, theory this paper presents an optimum design problem coming from mechanical engineering. This problem can be formulated as to determine the shape of the boundary of an elastic body in order to minimize the stress concentration near this boundary. This type of problems are solved by using an optimal control formulation in which the control is the boundary itself and the state is given by solving the equation of elasticity. The criterion, for instance the maximum value of stresses, is to be minimized by non linear programming methods. The paper presents a formulation of the problems and numerical solutions obtained on a specific example.

Keywords. Optimum design, elasticity, non linear programming.

INTRODUCTION

The theory of optimal control of systems governed by partial differential equations has been widely developed during the last ten years. An important range of applications of this theory consists in problems of optimum design. By an optimum design problem we mean a problem where the control is the geometry of the domain in which the boundary value problem is solved, see e.g. LIONS (1972). An abstract formulation of the problem is the following one.

Let Ω_α be a variable domain in R^n with

$$\partial\Omega_\alpha = \mathcal{C} \cup \Gamma_\alpha$$

where Γ_α is the variable part of the boundary which depends on a family of parameters $\alpha \in \mathcal{A}$, $\mathcal{A}$ being a space to be defined (see Fig.1).

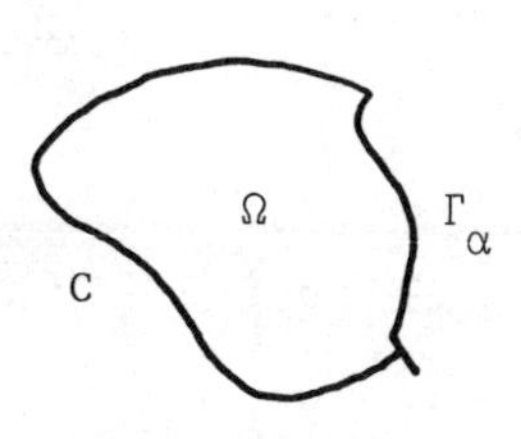

Figure 1.

Let u_α be the solution of the partial differential equation

$$Au_\alpha = f \quad \text{in } \Omega_\alpha \tag{1}$$

$$Bu_\alpha = g \quad \text{on } \partial\Omega_\alpha \tag{2}$$

Then for any value of α it is possible to define a functional $\mathcal{J}:\mathcal{A} \to R$:

$$\mathcal{J}(\alpha) = \Phi(u_\alpha) \quad . \tag{3}$$

The optimum design problem consists in

$$\min_{\alpha \in \mathcal{A}} \mathcal{J}(\alpha) \tag{4}$$

The space $\mathcal{A}$ is a space of parameters which determine the shape of Γ_α and the problem consists in minimizing $\mathcal{J}$ by choosing the shape of Γ_α.

The difficulties of this type of problems are both theoretical and practical. From a theoretical point of view it is necessary to settle a convenient functional framework in order to give a sense to problem (1) (2) (3) (4) and to obtain first order necessary conditions for optimality, concerning this point we refer to PIRONNEAU (1976) and MURAT-SIMON (1974).

From a practical point of view this problem requires a very efficient minimization method It must be noticed that the computation of $\mathcal{J}(\alpha)$ requires one solution of the boundary value problem (1) (2), and because of the change of domain at each iteration this represents a great computational effort. All these aspects of the problem will be detailed in the sequel on a specific example.

A MECHANICAL EXAMPLE

Compressing stages of engine, in aircrafts, have fitted disks with which blades are fitted. The centrifugal force tends to pull out the bases of blades and there appear very large stresses in the disk near the boundary of alveoles. (see Fig. 2).

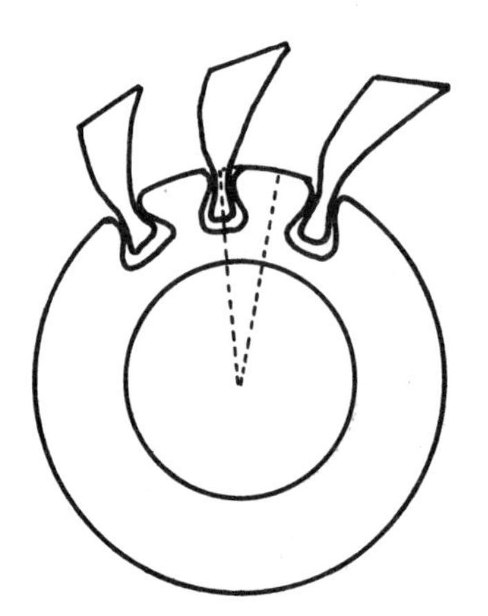

Figure 2.

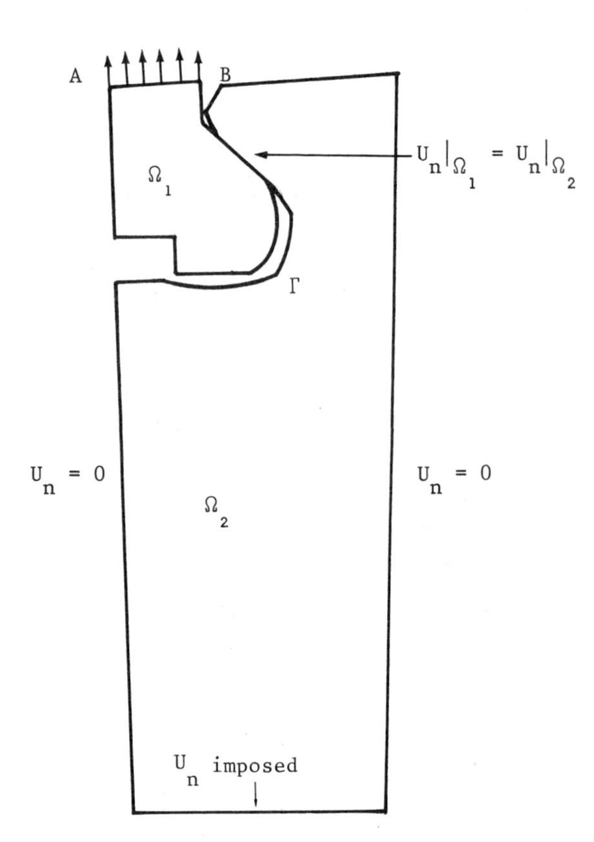

Figure 3.

The problem consists in optimizing the shape of the alveole in order to minimize a function of stresses. The domain is pictured in Fig. 10 where Ω_2 denotes a part of the disk (the structure of the disk is periodic) and Ω_1 is the basis of the blade. Boundary conditions are given on Fig. 3. In particular an imposed force is applied on the boundary AB. The shape of the boundary Γ (which is the part of the alveole with which the blade is fitted) is to be optimized. As we can neglect the effect of finite thickness of the disk, a two dimensional problem may be considered. The material is assumed linearly elastic, homogeneous and isotropic. Plane strain is assumed to calculate stresses.

The boundary Γ being given, equation of displacements $u = (u_x, u_y)$ is given by the virtual works principle :

$$\int_{\Omega_1 \cup \Omega_2} \sigma^t \tilde{\varepsilon} \, dx = \int_{\Omega_1 \cup \Omega_2} \tilde{u}^t F \, dx + \int_A^B \tilde{u}^t T \, ds \quad (5)$$

where $\tilde{\varepsilon}$ is the strain vector associated with the virtual displacement $\tilde{u}$ (x^t denotes the transpose of x), F is a distributed force (centrifugal) and T is the traction due to the blade.

To the solution u of (5) corresponds the stress σ

$$\sigma^t = (\sigma_x, \sigma_y, \tau_{xy})$$

and the strain

$$\varepsilon^t = (\varepsilon_x, \varepsilon_y, \gamma_{xy})$$

where

$$\varepsilon_x = \frac{\partial u_x}{\partial x} \qquad \varepsilon_y = \frac{\partial u_y}{\partial y} \qquad \gamma_{xy} = \frac{\partial u_x}{\partial y} + \frac{\partial u_y}{\partial x}$$

The elastic law of the medium gives

$$\sigma = D \, \varepsilon$$

where D is the 3 x 3 matrix

$$D = \begin{bmatrix} d_1 & d_2 & 0 \\ d_2 & d_1 & 0 \\ 0 & 0 & d_3 \end{bmatrix}$$

with

$$d_1 = \frac{E(1-\nu)}{(1+\nu)(1-2\nu)} \quad d_2 = \frac{E\nu}{(1+\nu)(1-2\nu)} \quad d_3 = \frac{E}{2(1+\nu)}$$

(E is the Young's modulus and ν the Poisson's coefficient).
It is classical that the variational equation (5) has a unique solution belonging to a subspace of the Hilbert space

$$V = [H^1(\Omega_1 \cup \Omega_2)]^2$$

(cf. for instance ODEN-REDDY (1976)).

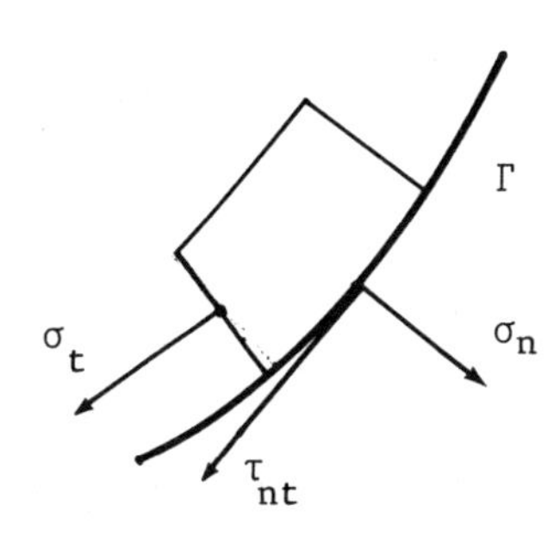

Figure 4.

On the boundary Γ the normal stress σ_n and shear stress τ_{nt} vanish (cf. Fig. 4).
From a mechanical point of view the main problem is to avoid appearance of cracks on the boundary. The cause of cracks is the stresses concentration σ_t near Γ. A crude way to solve this problem is to minimize a function of σ_t.

Then a natural criterion is the following

$$\mathcal{J}(\Gamma) = \max_{\gamma \in \Gamma} |\sigma_t(\alpha)|. \quad (6)$$

Nevertheless in order to avoid the dificulties of "min-max" problems we have also considered a "smoothed" version of (6) :

$$\mathcal{J}_\rho(\Gamma) = \int_\Gamma |\sigma_t(\alpha)|^{2\rho} \, d\gamma \quad (7)$$

which is clearly an approximation of (6) for ρ large enough.

Furthermore additional restrictions on the shape of Γ must be considered for several reasons. For instance the convexity of Γ must be constant in order to avoid difficulties for manufacturing. These points will be detailed below.

PARAMETRIZATION

The general approach is to combine the finite elements approximation of problem (5) and the parametrization of the variable boundary Γ. This has been done by many authors (see for instance MORICE (1975), PIRONNEAU (1976)).

Here we present a very simple method which reduces the number of unknowns. If we assume that the initial guess of Γ_α is not too far from the optimum it is possible to associate to $\{\Gamma_\alpha\}_\alpha$ a family of fixed curves $\{\Delta_i\}_{i=1}^p$ nearly orthogonal to $\{\Gamma_\alpha\}$. Theses curves will be called "meridians" in the sequel.

Then the boundary Γ_α will be determined by a finite set of parameters $\{\alpha_i\}_{i=1}^p$, where α_i is the curvilinear abscissa on Δ_i of the intersection point D_i of Δ_i with Γ_α. The finite elements approximation is then based on a moving mesh which uses the moving boundary nodes D_i $i = 1, 2, \ldots, p$ (see Fig. 8).

Nevertheless in order to avoid a complete modification of the finite element mesh for any change in the position of $\{D_i\}$ it is necessary to divide Ω_α into two parts (see Fig.9): a fixed part where the mesh is fixed and a moving part in which the mesh depends on the position of Γ_α.

The displacement of internal nodes in the moving part is given by simple rules. For instance the curvilinear abscissa of an internal node on Δ_i is an affine function of α_i.

FINITE DIMENSIONAL PROBLEM

Using the classical quadratic serendipity elements, see ZIENKIEWICZ (1971), associated to the mesh of Fig. 15, equation (5) is written in the following form :

$$K(\alpha) u_\alpha = f(\alpha)$$

where $K(\alpha)$ is an $r \times r$ matrix (stiffness matrix) and $u_\alpha \in R^r$.
Criterion (7) is written as :

$$\mathcal{J}_\rho(\alpha) = \sum_j |c_j^t(\alpha) \cdot u_\alpha|^{2\rho} d\gamma \quad (9)$$

where $c_j(\alpha) \in R^r$, $j = 1, 2, \ldots, q$.

The main step is to calculate the gradient of $\mathcal{J}_\rho(\alpha)$. This can be done by using an adjoint state which is usual in optimal control theory. We have :

$$\frac{d\mathcal{J}_\rho(\alpha)}{d\alpha} \cdot \tilde{\alpha} = 2\rho \sum_j |c_j^t(\alpha) u_\alpha|^{2\rho-1} [c_j^t \tilde{u}_\alpha + \frac{dc_j^t}{d\alpha} \tilde{\alpha} u_\alpha] \quad (10)$$

where $\tilde{u}_\alpha$ is the solution of the linearized problem :

$$\frac{dK(\alpha)}{d\alpha}\,\tilde{\alpha} u_\alpha + K(\alpha)\tilde{u}_\alpha = \frac{df}{d\alpha}\cdot\tilde{\alpha}.$$

The adjoint state $p(\alpha)$ is defined by :

$$K^t(\alpha)p = \sum_j |c_j^t(\alpha)u_\alpha|^{2\rho-1} c_j \quad . \quad (11)$$

Then :

$$p^t K(\alpha)u = \sum_j |c_j^t(\alpha)u_\alpha|^{2\rho-1} c_j^t u_\alpha$$

$$= p^t\left[\frac{df}{d\alpha}\tilde{\alpha} - \frac{dK(\alpha)}{d\alpha}\tilde{\alpha}\tilde{u}_\alpha\right].$$

Using this last expression in (10) we get :

$$\frac{d\mathcal{J}_\rho(\alpha)}{d\alpha}\tilde{\alpha} = 2\rho\; p^t\left[\frac{df}{d\alpha}\tilde{\alpha} - \frac{dK}{d\alpha}\,\alpha\tilde{u}_\alpha\right]$$

$$+ 2\rho \quad |c_j^t(\alpha)u_\alpha|^{2\rho-1}\frac{dc_j^t(\alpha)}{d\alpha}\,\tilde{\alpha}u_\alpha$$

In particular the ith component of the gradient is :

$$\frac{\partial \mathcal{J}_\rho(\alpha)}{\partial\alpha_i} = 2\rho\; p^t\frac{\partial f}{\partial\alpha_i} - 2\rho\; u_\alpha^t\; p^t\frac{\partial K(\alpha)}{\partial\alpha_i}$$

$$+ 2\rho \quad |c_j^t(\alpha)u_\alpha|^{2\rho-1} u_\alpha^t\frac{\partial c_j^t(\alpha)}{\partial\alpha_i} \qquad (12)$$

The gradient is given by one solution of the adjoint state equation (11). The crucial point is to calculate carefully the coefficients :

$$\frac{\partial K_{lm}}{\partial\alpha_i}$$

which give the derivative of the stiffness matrix coefficient K_{lm} with respect to the design parameter α_i.

THE NON LINEAR PROGRAMMING PROBLEM

As we have mentioned earlier we have to introduce additional constraints on the variable $\{\alpha_i\}_{i=1}^{p}$ in order to keep a constant convexity of the boundary. For instance the boundary Γ pictured on Fig. 5 is unfeasible, we want to obtain a shape as on Fig. 6.

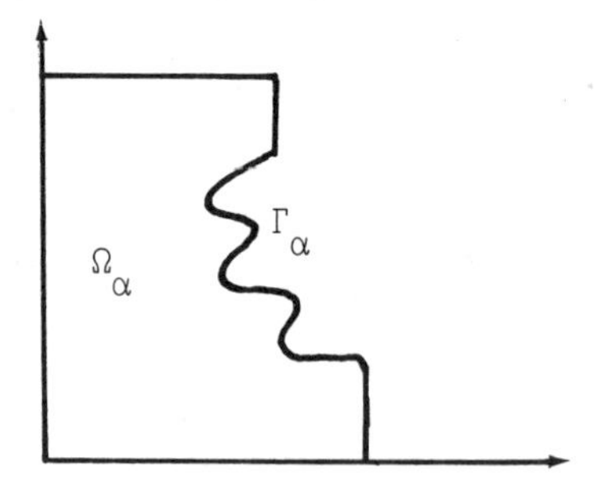

Figure 5.

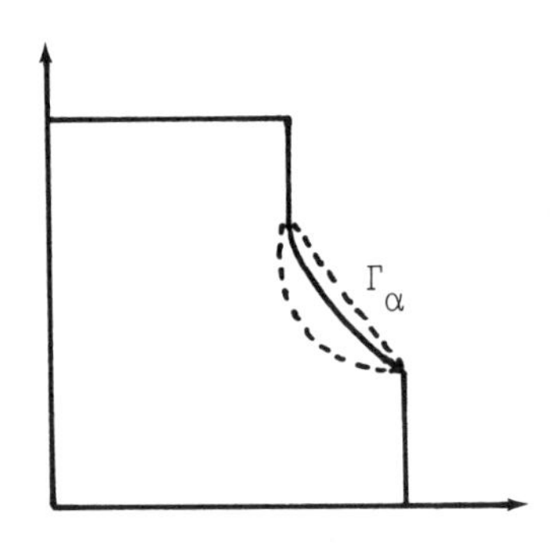

Figure 6.

Let us represent on Fig. 7 a part of the boundary where D_i is a moving node associated with α_i and θ_i is the oriented angle.

$$\theta_i = (\vec{oy}, \overrightarrow{D_i D_{i+1}})$$

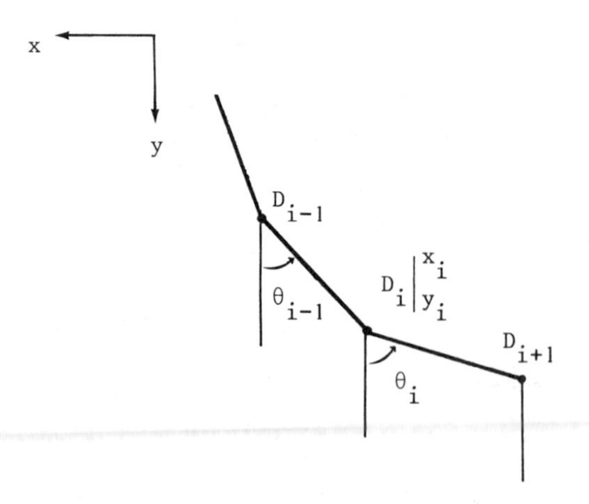

Figure 7.

The condition of constant convexity imposes

$$\theta_i \geq \theta_{i+1}$$

or, if $-\frac{\pi}{2} < \theta_i < \frac{\pi}{2}$,

$$\mathrm{tg}\ \theta_i \geq \mathrm{tg}\ \theta_{i+1} .$$

This condition can be written obviously as a function of the coordinates (x_i, y_i) of D_i

$$\frac{x_i - x_{i-1}}{y_i - y_{i-1}} \geq \frac{x_{i+1} - x_i}{y_{i+1} - y_i}$$

(in the case of Fig.7).

This last inequality can be rewritten by using the design parameters α_i's and this gives the following inequality

$$C_i(\alpha) = \bar{a}_i + \bar{b}_i \alpha_{i-1} + \bar{c}_i \alpha_i + \bar{d}_i \alpha_{i+1} + \bar{e}_i \alpha_{i-1} \alpha_i + \bar{f}_i \alpha_i \alpha_{i+1} + \bar{g}_i \alpha_{i-1} \alpha_{i+1} < 0 \quad (13)$$

where $\bar{a}_i, \bar{b}_i, \ldots, \bar{g}_i$ are constants, $i = 1,2, \ldots, p$.

Furthermore, as we know roughly the position of the optimal solution it is possible to impose restrictions on the values of α_i's

$$\alpha_i^m \leq \alpha_i \leq \alpha_i^M \qquad i = 1, 2, \ldots, p \quad (14)$$

where α_i^m et α_i^M are given.

OPTIMIZATION METHODS

The nonlinear programming problem consists in minimizing (9) under constraints (13) and (14). This problem is non convex but it is possible to use a (generalized) reduced gradient method. The basis of the method is that if the constraint $C_k(\alpha)$ is active, i.e.

$$C_k(\alpha) = 0, \quad (15)$$

then it is possible to express α_k in function of α_j for $j = k-1, k+1$. This allows to reduce the number of variables by eliminating α_k. This element is quite simple to perform. Of course this method requires some "pivoting" steps when one constraint (14) is active.

As we have mentioned earlier the criterion (9) is a "smoothed" form of criterion (6) which takes the form, in the finite dimensional case :

$$\mathcal{J}(\alpha) = \max_j c_j^t(\alpha)\, u_\alpha . \quad (16)$$

As it is classical the minimization of (16) can be transformed in the new problem;

$$\min_{\alpha, \beta} \beta \quad (17)$$

$$c_j^t(\alpha)\, u_\alpha \leq \beta \qquad j = 1, 2, \ldots, q \quad (18)$$

with the same other constraints (13) and (14). In this case it is not easy to use the same method as previously because of constraints (18) which are not simple at all.

We have used for this last problem the method proposed by several authors like BIGGS (1975), HAN (1976) and POWELL (1978), which is based on augmented Lagrangian.

NUMERICAL RESULTS

The initial domain before optimization is given on Fig. 10. The result after optimization (reduced gradient) of criterion $\mathcal{J}_\rho$, (9) with $\rho = 8$, is shown on fig. 11. The fig. 12 represents the same results after optimization of the criterion (16) $\mathcal{J}(\alpha)$ by the Powell's method.
Table 1 gives a comparizon of convergences of the two algorithmes.

CRITERION	NUMBER OF ITERATIONS	MAXIMUM STRESS VALUE
J_ρ	15	43.71
J_∞	3	37.58

(initial value of maximum stress : 50.06)

Table 1

Figures 13 and 14 show the different results obtained by minimizing $\mathcal{J}_\rho$ and $\mathcal{J}$.

CONCLUDING REMARKS

The efficiency of the Powell's method is patent. It seems to be very well adapted to highly non linear problems.

This fact is very important from pratical point of view. In optimum design problems the calculation of one value of the criterion and/or the gradient is very expensive in computer time and the feasability of the method is closely related to the efficiency of the optimization method. In the near future a lot of optimum design problems will be solved numerically, because this field seems to be very promising in mechanical engineering, electrical engineering, etc... (see for instance, PIRONNEAU (1973)(1974), MORICE (1974), KELLY and al. (1977), MIDDLETON-OWEN (1977)).

For the case of distributed systems this type of problem has given new interesting extensions of the classical control theory.

REFERENCES

Biggs, . C. (1975). Constrained optimization using recursive quadratic programming. L.C.W. DIXON and G.P. SZEGO Eds. Toward Global Optimization. North-Holland, Amsterdam.

Han, S.P. (1976). Superlinearly convergent variable metric algorithms. Math. Programming, 11, pp. 263-282.

Kelly, D.W., Morris, A.J., Bartholomew, P., Stafford, R.D. (1977). A review of techniques for automatic structural design. Comp. Methods in Applied Mechanics and Engineering, 12, pp. 219-242.

Lions, J.L., (1972). Some aspects of the optimal control of distributed parameter systems. Regional Conference Series in Applied Math., SIAM n°6.

Middleton, J., Owen, D.R.J. (1977). Automated design optimization to minimize stearing stress in axisymetric pressure vessels. Nuclear Engineering and Design, 44, pp. 357-366.

Morice, P. (1975). Une méthode d'optimisation de forme. Proc. Conference IFIP-IRIA 1974. Springer-Verlag, pp. 454-467.

Murat, F., Simon, J. (1977). Etude de quelques problèmes d'optimisation. Publication Univ. Paris VI.

Oden, J.T., Reddy, J.N. (1976). Variational methods in theoretical mechanics. Universitext, Sringer-Verlag, Berlin.

Pironneau, O., (1973). On optimum profiles in stoker flow. J. Fluid Mech., Vol. 59, pp. 117-128.

Pironneau, O. (1974). On optimum design in fluid mechanics. J. Fluid Mech., Vol. 64, pp. 97-114.

Pironneau, O., (1976). Ph. D. Thesis, Paris VI.

Powell, M.J.D., (1978). Algorithms for non linear constraints that use Lagrangian functions. Math. Programming, 14, pp. 224-248.

Zienkiewicz, O.C. (1971). The finite element method. Mc. Graw-Hill, London.

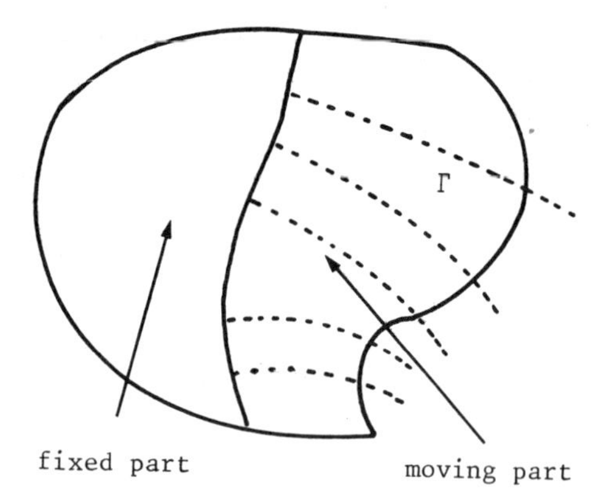

Figure 9.

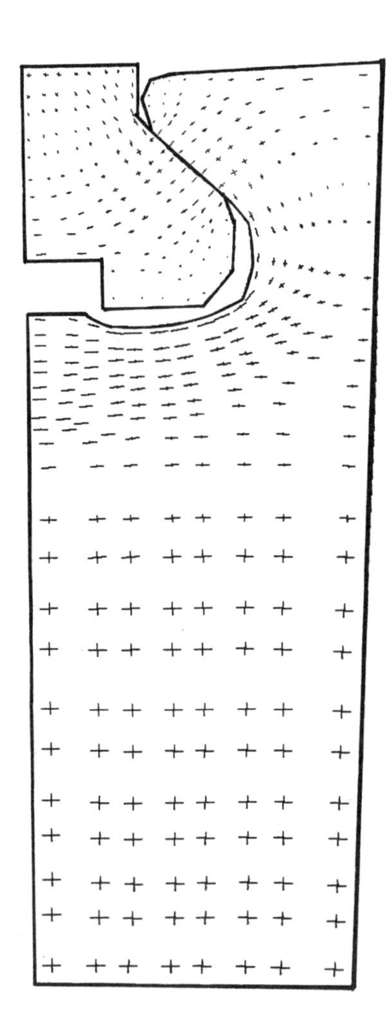

Figure 10.

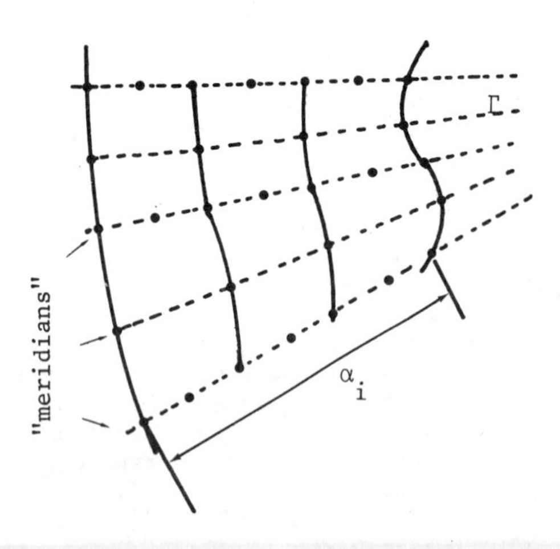

Figure 8.

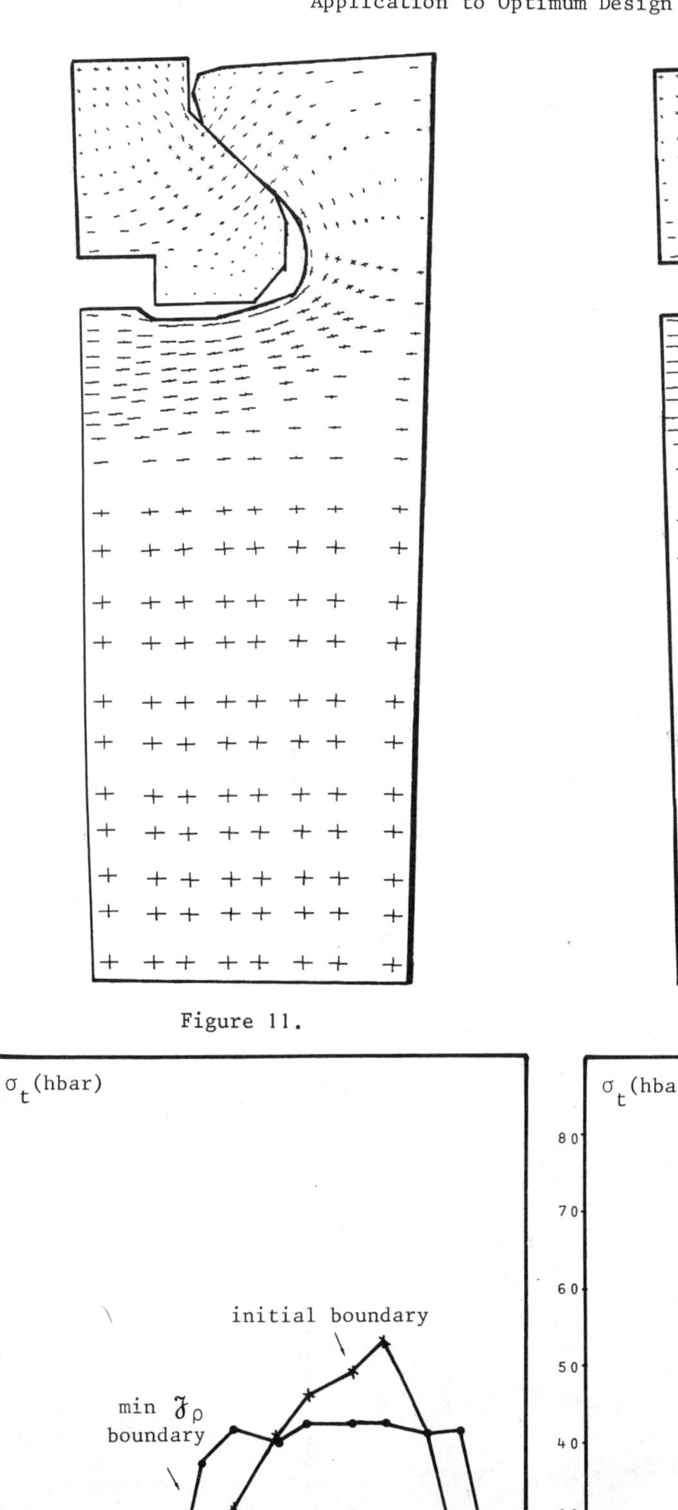

Figure 11.

Figure 12.

Figure 13.

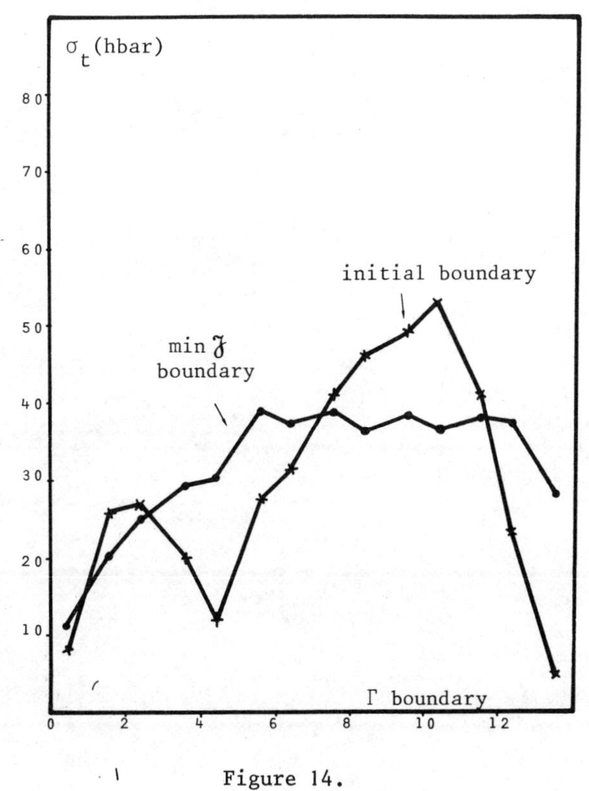

Figure 14.

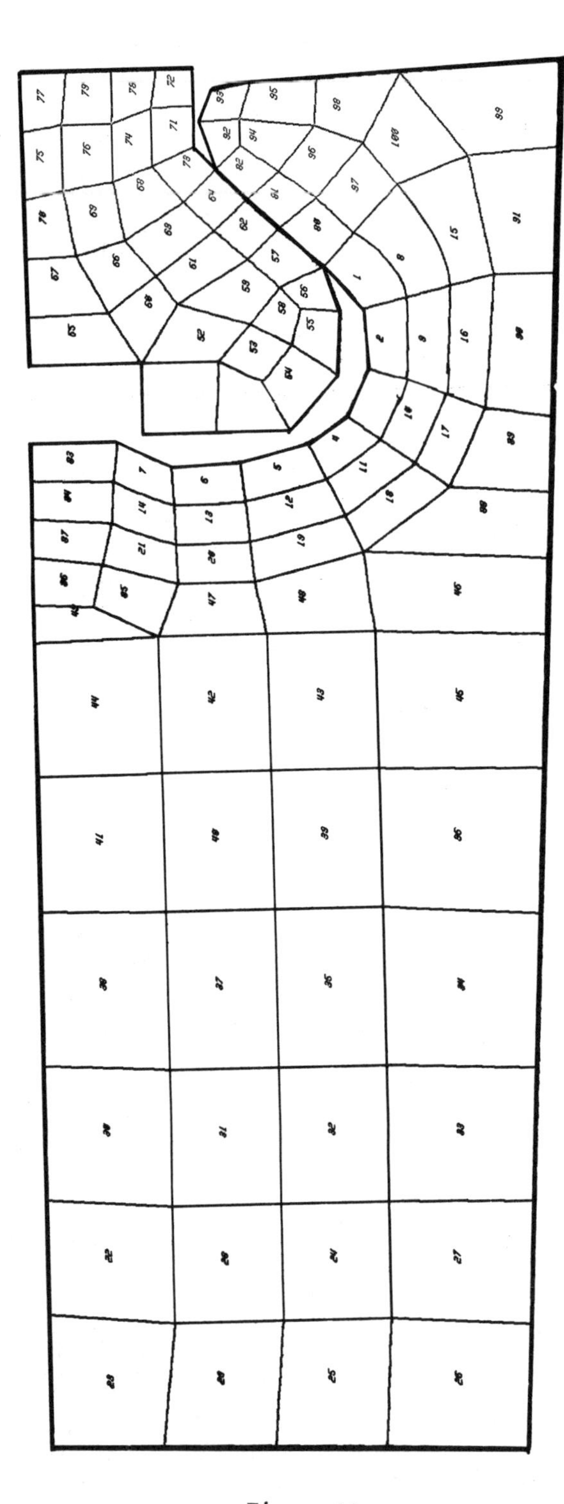

Figure 15.

NONLINEAR REGULATOR DESIGN FOR MAGNETIC SUSPENSION

T. L. Vincent

Aerospace and Mechanical Engineering, University of Arizona, Tucson 85721, USA

Abstract. Through proper design, it is possible to build an electromagnetic bearing system using a ferromagnetic rotor, such that there exists an equilibrium position in the magnetostatic field which is unstable only in the axial direction. In order to achieve axial stability, a regulator may be employed to vary the current in the coils whenever the rotor is displaced from the equilibrium position. The idea is to vary the current in such a way so that the resultant change in the magnetic field produces a restoring force to the rotor. The problem of concern here is the design of a regulator to control the magnetic field so that the rotor will remain in the vicinity of the equilibrium point in the presence of external forces (noise).

Recent magnetic suspension systems have been designed using a regulator based on linear state variable feedback. It is shown here that asymptotic stability can not be maintained for the system in the presence of noise if linear feedback control is used. The effect of noise on the system is vividly demonstrated by using qualitative methods to find the boundaries of the reachable set under noise. Linear state variable feedback can maintain partial controllability of the system to some neighborhood of the equilibrium point provided a simple inequality relationship between the feedback parameters and the bounds on the noise level is maintained. Finally, it is shown that a game theoretic approach may be used for a nonlinear regulator design which will result in an asymptotically stable system in the presence of bounded noise.

Keywords. Nonlinear control systems; Stability; Controllability; Feedback; Differential games.

INTRODUCTION

The objective is to suspend a ferromagnetic rotor in a magnetic field. In a static inverse-square force field there will always be at least one statically unstable coordinate direction (Earnshaw, 1842). However, by properly varying the magnetic field, stable suspension is possible. For example, Sabnis, Dendy, and Schmitt (1975) discussed the design and fabrication of a large magnetically suspended momentum wheel. The bearing was designed in such a way that passive magneto static stability is obtained for a rotor in the radial direction. The static instability in the axial direction was successfully controlled by a position feedback lead compensator. More recent systems developed by Sperry Rand Corporation uses velocity as well as position information for control.

A simple model for a magnetically suspended rotor is obtained by assuming one dimensional motion along an unstable axial direction as shown in Fig. 1. It is further assumed that an inverse-square force field exists between the gaps. Thus a rotor of length 2ℓ, where center of mass is displaced by a distance y from the centerline between the poles, will experience a force of attraction given by

$$f_1 = K_1/(K_2 + z_1)^2 \tag{1}$$

to the upper pole and a force of attraction given by

$$f_1 = K_1/(K_2 + z_2)^2 \tag{2}$$

to the lower pole, where z_1 is the gap between the top of the rotor and the upper pole, z_2 is the gap between the bottom of the rotor and the lower pole, and K_1 and K_2 are constants associated with the magnetic system. Let 2d be the distance between the poles then

$$z_1 = d - \ell - y \tag{3}$$

and

$$z_2 = d - \ell + y . \tag{4}$$

Finally by letting $K_3 = K_2 + d - \ell$, the net force toward the upper pole is given by

$$f_1 - f_2 = \frac{K_1}{(K_3 - y)^2} - \frac{K_1}{(K_3 + y)^2} \tag{5}$$

or equivalently

$$f_1 - f_2 = 4K_1K_3y/(K_3^2 - y^2)^2 . \tag{6}$$

In terms of a first order expansion about $y = 0$, the force difference may be approximated by

$$f_1 - f_2 = K_u y \tag{7}$$

where $K_u = 4K_1/K_3^3$ represents the "stiffness" of the unstable rotor. In addition to this static magnetic force it is assumed that an additional force, r, can be created by applying current to the control coils and that external forces (noise), s, may act directly on the rotor. Hence the axial acceleration of the rotor is determined from

$$M\frac{d^2y}{d\tau^2} = K_u y + r + s \tag{8}$$

where M is the mass of the rotor and τ is time. It is further assumed that the control force r and external forces s are bounded,

$$|r| \leq r_m \tag{9}$$

$$|s| \leq s_m \tag{10}$$

with $0 \leq s_m < r_m$.

By letting $y_1 = y$, $y_2 = \dot{y}$, and $\alpha^2 = K_u/M$, equation (8) may be put into a convenient state variable format.

$$dy_1/d\tau = y_2 \tag{11}$$

$$dy_2/d\tau = \alpha^2 y_1 + \frac{r}{M} + \frac{s}{M} . \tag{12}$$

The equations are further simplified by using the nondimensional variables defined by

$$t = \alpha\tau \tag{13}$$

$$x_1 = (\alpha^2 M/r_m)y_1 \tag{14}$$

$$x_2 = (\alpha M/r_m)y_2 \tag{15}$$

$$u = r/r_m \tag{16}$$

$$v = s/s_m \tag{17}$$

$$\beta = s_m/r_m . \tag{18}$$

In terms of these nondimensional variables equations (11) and (12) become

$$\dot{x}_1 = x_2 \tag{19}$$

$$\dot{x}_2 = x_1 + u + \beta v \tag{20}$$

where the dot denotes differentiation with respect to non-dimensional time. Note that

$$|u| \leq 1, \ |v| \leq 1, \ 0 \leq \beta < 1. \tag{21}$$

The instability of the axial motion without control or noise inputs ($u = v = 0$) is easily demonstrated from (19) and (20). Eliminating time and integrating, results in hyperbolic state space trajectories

$$x_1^2 - x_2^2 = \text{constant} \tag{22}$$

as illustrated in Fig. 2 for various values of the constant.

LINEAR FEEDBACK CONTROL

Let $x = [x_1, x_2]^T$, then in terms of matrix notation equations (19) and (20) may be written as

$$\dot{x} = Ax + B(u + \beta b) \tag{23}$$

where $A = \begin{bmatrix} 0 & 1 \\ 1 & 0 \end{bmatrix}$ and $B = \begin{bmatrix} 0 \\ 1 \end{bmatrix}$. With $\beta = 0$, the system (23) is "completely controllable" (Brogan, 1974) that is the matrix $[B \vdots AB]$ has rank 2. This means that there exists positive feedback parameters h_1 and h_2 such that a linear feedback control law given by

$$u = -h_1x_1 - h_2x_2 \tag{24}$$

will guarantee asymptotic stability from any point in the state space for the system (23) provided $\beta = 0$. However for any finite h_1, h_2 the control law given by (24) will violate $|u| \leq 1$ for most points. Clearly imposing the physical constraint $|u| \leq 1$ is not compatible with the usual notions of complete controllability. For example, with $\beta = 0$ and u given by (24), the characteristic equation for (23) is given by

$$\lambda^2 + \lambda h_2 + (h_1 - 1) = 0 , \tag{25}$$

So that in addition to $h_2 > 0$ asymptotic

stability requires $h_1 > 1$. Complete controllability implies arbitrary pole placement, so if the roots of (25) are given by

$$\lambda_1 = -a_1 - ib \quad (26)$$

$$\lambda_2 = -a_2 + ib \quad (27)$$

where $a,b > 0$, $a_1 = a_2$ if $b \neq 0$, it follows that

$$h_1 = a_1 + a_2 \quad (28)$$

$$h_2 = 1 + a_1 a_2 + b^2 \ . \quad (29)$$

Suppose it was desired to have the roots at $\lambda_1 = -1$, $\lambda_2 = -2$, it then follows from (28) and (29) that $h_1 = h_2 = 3$. Figure 3 illustrates state space trajectories for the system (23) with $\beta = 0$ under linear feedback control (24) with $h_1 = h_2 = 3$. The control violates $|u| \leq 1$ outside the parallel lines defined by $|3x_1 + 3x_2| > 1$ as indicated on the figure. However, asymptotic stability is maintained within the region defined by $|3x_1 + 3x_2| \leq 1$ since all trajectories starting from points on $|3x_1 + 3x_2| = 1$ are directed into the region. This is easily verified by examining dx_2/dx_1 as determined from (19) and (20). It is evident that there is a larger stability region for this system which satisfies $|u| \leq 1$ than the one just obtained. Clearly different results will be obtained by using different values for the feedback parameters h_1 and h_2. It should be noted again that imposing $|u| \leq 1$ greatly restricts pole placement if one is seeking a "large" stability region. It will be shortly demonstrated that for this problem under linear feedback control, with $\beta = 0$, the largest stability region which satisfies $|u| \leq 1$ is obtained when $h_1 = h_2$. The general problem of determining the "largest" stability region will be discussed shortly under nonlinear control.

The effect of noise on the system under linear feedback control may be examined by determining the set of points reachable from the origin under v control for the system (23) with u control as given by (24). That is for the system

$$\dot{x}_1 = x_2 \quad (30)$$

$$\dot{x}_2 = (1 - h_1)x_1 - h_2 x_2 + \beta v \quad (31)$$

where $v \leq 1$. In order to satisfy $|u| \leq 1$, the state space is restricted to points which satisfy $|h_1 x_1 + h_2 x_2| \leq 1$. Trajectories which lie in the boundary of the v reachable set about the origin may be generated from the necessary condition that v maximizes

$$H = \lambda_1 x_2 + \lambda_2[(1 - h_1)x_1 - h_2 x_2 + \beta v] \quad (32)$$

on the boundary of the v reachable set (Grantham and Vincent, 1975) where λ_1 and λ_2 are determined from the adjoint equations

$$\dot{\lambda}_1 = \lambda_2(h_1 - 1) \quad (33)$$

$$\dot{\lambda}_2 = -\lambda_1 + \lambda_2 h_2 \quad (34)$$

such that $H = 0$. Since

$$\frac{\partial H}{\partial v} = \lambda_2 \beta \quad (35)$$

and since $\lambda_2 \equiv 0 \Rightarrow \lambda_1 \equiv 0$, the v control on the boundary of the v reachable set must be bang-bang. Consider the case where $h_1 = h_2 = 3$. The solution for λ_2 as determined from (33) and (34) is given by

$$\lambda_2 = C_1 e^t + C_2 e^{2t} \quad (36)$$

where C_1 and C_2 are constants determined by initial conditions. According to (36), λ_2 can change sign at most once which means that after some period of time the control v is a constant plus or minus one. Under this constant control, the system, governed by (30) and (31), has a stable equilibrium point given by

$$x_2 = 0, \ x_1 = \beta v/2 \ . \quad (37)$$

It follows then that two points on the boundary of the v reachable set are given by $(\beta/2, 0)$ and $(-\beta/2, 0)$. Consider starting the system at the point $(\beta/2, 0)$. It follows that $\lambda_2(0) > 0$ implies $v = 1$ which satisfies $H = 0$. Under these initial conditions, the system remains at $(\beta/2, 0)$ until such time that λ_2 changes sign. If there is a sign change at time $t_s > 0$, then λ_2 remains negative for $t > t_s$ according to (36) and the system will move to equilibrium point $(-\beta/2, 0)$ under the constant control $v = -1$. Thus a candidate for the boundary of the v reachable set is easily determined by simply integrating the system forward from the points $(\beta/2, 0)$ and $(-\beta/2, 0)$ under appropriate constant control.

Figure 4 illustrates the v reachable set for the system (30) - (31) with $h_1 = h_2 = 3$ and $\beta = .2$. Clearly, under a "worst case" situation, it is possible for the noise input, v, to keep the system on the boundary illustrated. In general, however, it is expected the system would lie somewhere within these boundaries. In any event, linear feedback control with finite gains h_1 and h_2 can not maintain asymptotic stability to the origin in the presence of noisy inputs. For the case illustrated, it is clear that the linear feedback control results in partial stability or ultimate confinement. That is any starting point satisfying $|3x_1 + 3x_2| \leq 1$ but outside the v reachable set will be driven to and maintained within or on the boundary of the v reachable set.

NONLINEAR FEEDBACK CONTROL

There are two qualitative aspects of the control system given by (19) and (20) which need to be sorted out. One has to do with the "size" of the stability region and the other has to do with the possible loss of asymptotic stability with noise inputs. By adding an additional feedback term to the usual linear feedback control law (Gutman and Leitmann, 1976; Leitmann, 1979) have shown that asymptotic stability can be guaranteed for many linear systems. Their methods have been applied to the system under investigation as well as others for the purpose of comparing different approaches. The results of these investigations will be reported elsewhere. Among the number of possible ways of investigating the above mentioned qualitative aspects of control, the method ultimately used here is similar to one previously used by Ragade and Sarma (1967).

A naive approach (yet in the limit a good one, as will be shown) to nonlinear feedback control is to simply allow the linear feedback control law given by (24) to saturate

$$u = \begin{cases} -(h_1x_1 + h_2x_2) & \text{if } |h_1x_1 + h_2x_2| < 1 \\ -\text{sgn}(h_1x_1 + h_2x_2) & \text{if } |h_1x_1 + h_2x_2| \geq 1. \end{cases} \tag{38}$$

In this case, the constraint $|u| \leq 1$ is satisfied. The resulting stability region for the case where $\beta = 0$ may be estimated by simply integrating the system (19) and (20) backward in time under the control law (38) from various points in the neighborhood of the origin. The results for the case where $h_1 = h_2 = 3$ is illustrated in Fig. 5 with several typical trajectories drawn in forward time. The boundaries of the stability region shown are straight lines of slope 135° passing through ±1 on the x_1 axis. It is interesting to note that the same boundaries are obtained for all values of $h_1 = h_2 > 1$. If $h_1 \neq h_2$, then a smaller stability region is obtained as illustrated in Fig. 6.

With noise in the system, $\beta \neq 0$, a v reachable set about the origin may be obtained as before with u given by (38). The next step is to determine the set of points which are controllable to the v reachable set about the origin. The boundaries of this controllable set depend on β and may be obtained by using a modified game theoretic approach. The idea is to determine a guaranteed controllable set under the specified control law for u as given by (38). This can be done by seeking a "worst case" v strategy. That is "v" will use a strategy appropriate for a barrier in a game of kind (Isaacs, 1965). Trajectories which lie in the boundary of such a barrier (guaranteed controllable set under the specified control law for u) may be generated from the necessary condition that v maximize

$$H = \lambda_1x_2 + \lambda_2(x_1 + u(x_1, x_2) + \beta v) \tag{39}$$

on the boundary of the barrier where λ_1 and λ_2 are determined from the adjoint equations

$$\dot{\lambda}_1 = -\lambda_2(1 + \frac{\partial u}{\partial x_1}) \tag{40}$$

$$\dot{\lambda}_2 = -\lambda_1 - \lambda_2 \frac{\partial u}{\partial x_2} \tag{41}$$

such that $H = 0$. Again the v control must be bang-bang. Because of this, equilibrium solutions under constant control, $v = \pm 1$, must be points on the barrier. There are two possibilities at the equilibrium points, either $|u| < 1$ or not. If $|u| < 1$, then the equilibrium points are given by

$$x_1 = \pm\beta/(h_1 - 1),\ x_2 = 0 \tag{42}$$

which are identical to the equilibrium points for the v reachable set. If $|u| = 1$, then equilibrium points are given by

$$x_1 = -(u + \beta v),\ x_2 = 0\ . \tag{43}$$

Suppose $x_1 > 0$, then by (38), $u = -1$. Consider now the equilibrium solution corresponding to $v = -1$. If at this point, v is switched to $v = +1$, then $\dot{x}_2$ is positive. That is, the trajectory will be directed away from the origin. However, at the equilibrium solution corresponding to $v = +1$, if v is switched to $v = -1$, then $\dot{x}_2$ is negative with the trajectory directed toward the origin. Hence, only the latter equilibrium point is on the guaranteed controllable set boundary. Using the same argument for $x_1 < 0$ results in the following barrier equilibrium points

$$x_1 = \pm(1 - \beta),\ x_2 = 0\ . \tag{44}$$

Clearly the boundary of the guaranteed controllable set to the v reachable set must lie outside the boundary of the v reachable set. Thus, from (42) and (44), this requirement yields

$$(1 - \beta) > \beta/(h_1 - 1) \quad (45)$$

or equivalently (under the assumption $h_1 > 1$)

$$h_1 > 1 + \beta/(1 - \beta) . \quad (46)$$

If this condition is not satisfied, it is always possible to find a v control to completely destabilize the system. On a more positive note, if the condition is satisfied then the feedback control given by (38) has a guaranteed controllable set to the v reachable set as illustrated in Fig. 7 for the case of $h_1 = h_2 = 3$ and $\beta = .6$. For fixed values of $h_1 = h_2$, the effect of increasing β is to increase the size of the v reachable set, and at the same time squeeze in the outer boundary. Both effects are predicted by equations (42) and (44). It is when the two boundaries intersect, that the system can be completely destabilized by noise.

It is clear from (46) that as $\beta \to 1$, $h_1 \to \infty$ to maintain controllability. Note also from (42) that $h_1 \to \infty$ implies an infinitesimally small v reachable set. This suggests the desirability of using a feedback control law based on $h_1 = h_2 = \infty$. In terms of the previous saturation approach for handling $|u| \leq 1$, the control law becomes

$$u = -\text{sgn}(x_1 + x_2) \text{ if } (x_1 + x_2) \neq 0$$

$$u \in [-1, +1] \quad \text{if } (x_1 + x_2) = 0 . \quad (47)$$

Since the system given by (19) and (20) is set valued on the switching line, it is a generalized dynamical system. However for this special case it can be shown that solutions exist (Gutman and Leitmann, 1976).

A regulator based on this bang-bang control law does indeed result in an infinitesimally small v reachable set. That is under this control law the system defined by (19) - (20) is asymptotically stable to the origin, under all possible noisy inputs, in a region about the origin defined by

$$|x_1 + x_2| < 1 - \beta . \quad (48)$$

This is demonstrated by using a method of Liapunov (LaSalle and Lefschetz, 1961). A useful Liapunov function in this case if given by

$$V = (x_1 + x_2)^2 . \quad (49)$$

Thus

$$\dot{V} = 2(x_1 + x_2)(\dot{x}_1 + \dot{x}_2) \quad (50)$$

Substituting the system equations (19) - (20) with the control (47) yields

$$\dot{V} = 2(x_1 + x_2)^2 - 2(x_1 + x_2)\text{sgn}(x_1 + x_2) + 2\beta(x_1 + x_2)v \quad (51)$$

Imposing $\dot{V} < 0$ requires

$$(x_1 + x_2)^2 + \beta(x_1 + x_2)v < |x_1 + x_2| . \quad (52)$$

Clearly the worst case situation for v corresponds to $v = \text{sgn}(x_1 + x_2)$ in which case (52) reduced to (48). However, since $V = 0 \Leftrightarrow x_1 + x_2 = 0$, it has only been shown that the system is stable to the line $x_1 + x_2 = 0$.

To show asymptotic stability, let $y_2 = x_1 + x_2$ so that the system equations become

$$\dot{x}_1 = y_2 - x_1 \quad (53)$$

$$\dot{y}_2 = y_2 + u + \beta v \quad (54)$$

It has just been shown that y_2 is bounded and $y_2 \to 0$ asymptotically, hence from (53) $x_1 \to 0$ asymptotically. Hence, it is concluded the system is asymptotically stable to the origin (or at least stable to a small neighborhood about the origin, the size of which would depend on sampling and processing time required to implement the control law (47) in a real system) in the region defined by (48).

Even though use of (47) reestablishes asymptotic stability, it is not clear that the region defined by (48) is the largest possible stability region to the origin. It is of interest then, to compare yet another approach to nonlinear regulator design. Suppose it were known what set of points were controllable to the origin under all possible noise inputs to the contrary. Then from all such points, it may be possible to find strategies $u^*(\cdot)$ and $v^*(\cdot)$ which would produce a game theoretic saddle solution for the time to the origin. That is any other strategy $u(\cdot)$ played against $v^*(\cdot)$ which will result in a longer time to the origin and any other strategy $v(\cdot)$ played against $u(\cdot)^*$ will result in a shorter time to the origin.

For the system (19) and (20), necessary conditions to be satisfied by $u^*(\cdot)$ and

$v^*(\cdot)$ are that there exist nonzero multipliers $\lambda_1(t)$ and $\lambda_2(t)$ satisfying

$$\dot{\lambda}_1 = -\lambda_2 \qquad (55)$$

$$\dot{\lambda}_2 = -\lambda_1 \qquad (56)$$

such that

$$H = \lambda_1 x_2 + \lambda_2(x_1 + u + \beta v) \qquad (57)$$

takes on a maximum with respect to $u \in (-1,1)$ and a minimum with respect to $v \in (-1,1)$ at every point of the trajectory $x^*(t)$ generated by $u^*(\cdot)$ and $v^*(\cdot)$. Furthermore, $H(\lambda_1, \lambda_2, x^*, u^*, v^*) = 1$. Since $\partial H/\partial v = \beta \partial H/\partial u = \beta\lambda_2$ and since $\lambda_2 \equiv 0 \Rightarrow \dot{\lambda}_2 \equiv 0 \Rightarrow \lambda_1 \equiv 0$, both u and v control are bang-bang with opposite sign. It follows from (55) and (56), there can be at most one change of sign for λ_2 which implies at most one switch in the control sequence.

State space solutions are obtained by dividing (20) by (19), rearranging to obtain

$$x_2 dx_2 = x_1 dx_1 + (u + \beta v)dx_1 \qquad (58)$$

and integrating over a period of time for which u and v are constant to obtain

$$(x_1 + u + \beta v)^2 - x_2^2 = C + (u + \beta v)^2 \qquad (59)$$

where C is a constant of integration. Thus all trajectories are hyperbolic arcs. The complete state space plot is shown in Fig. 8. For completeness, trajectories outside the stability region are shown dashed. Note the switching arc corresponds to a solution through the origin with no change in sign for λ_2. It also corresponds to $C = 0$ in (59). The game theoretic solution for $u^*(\cdot)$, then, is to use $u = -1$ above the switching arc and $u = +1$ below the switching arc. Under this control law, the candidate for a guaranteed asymptotic controllable set are the points defined by $|x_1 + x_2| < 1 - \beta$ which is the same as the previous result.

The worst case for noise is when v control is just the opposite of the u control. It is easy to show through simulation, that if noise does not use the game theoretic strategy $v^*(\cdot)$, then the system remains asymptotically stable within $|x_1 + x_2| < 1 - \beta$ and moves more quickly toward the origin.

COMMENTS

It is clear, both from the limiting saturation approach and the game theoretic approach that a regulator based on bang-bang control is more desirable than a regulator based on linear state variable feedback control from the standpoint of being able to maintain asymptotic stability in the presence of noise. If one insists on linear state variable feedback, then the type of qualitative analysis presented here on determining the v reachable set is useful in choosing the magnitude of the feedback gains.

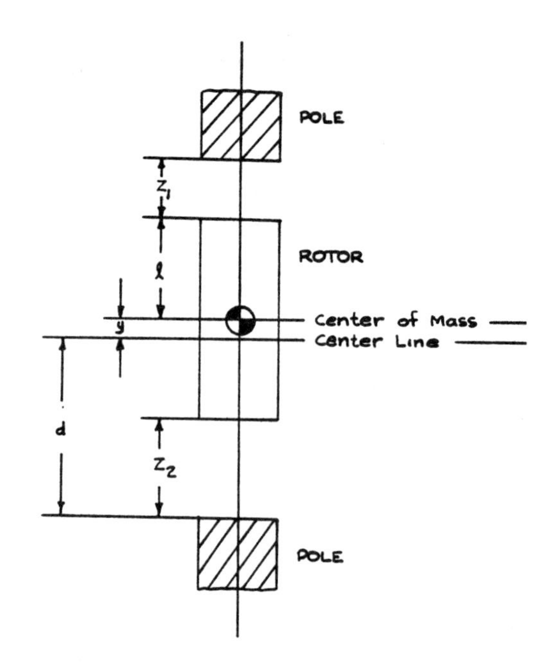

Fig. 1. Magnetically suspended Rotor.

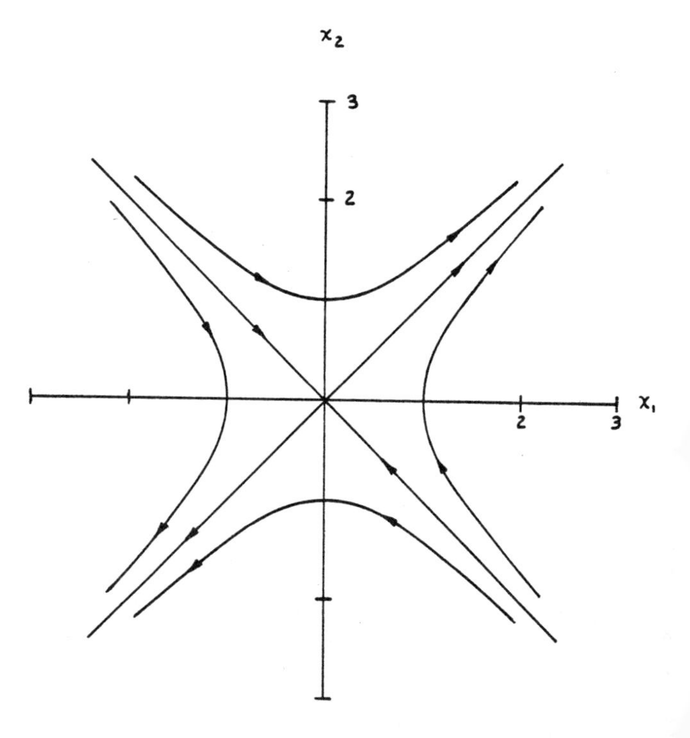

Fig. 2. Instability of motion under no control.

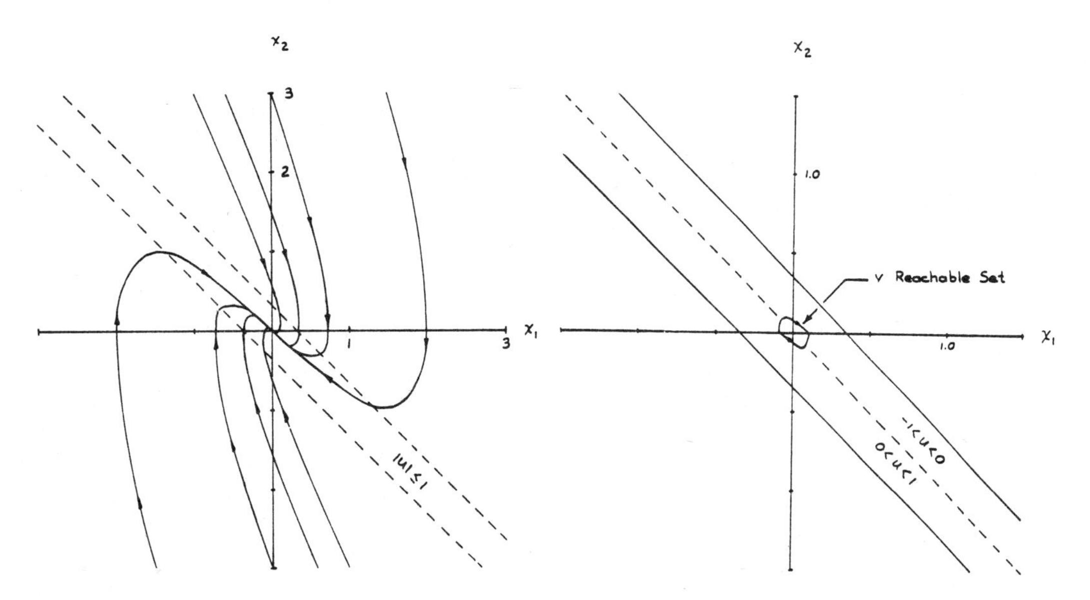

Fig. 3. Stable motion under unbounded linear state variable feedback.

Fig. 4. The v reachable set under linear state variable feedback.

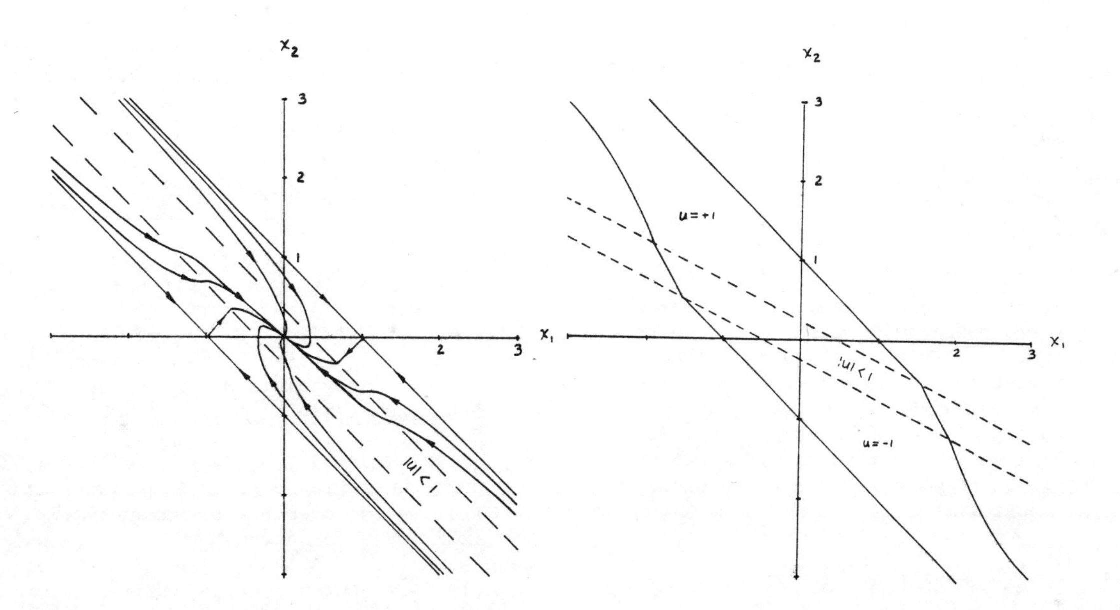

Fig. 5. Region of stability under modified linear state variable feedback with $h_1 = h_2$.

Fig. 6. Region of stability under modified linear state variable feedback with $h_1 \neq h_2$.

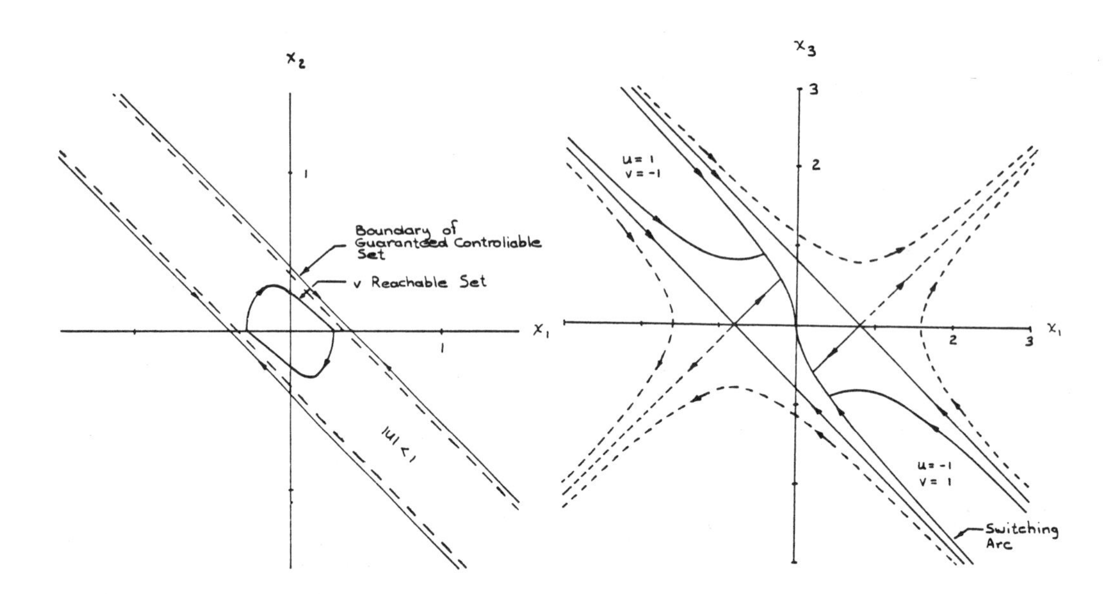

Fig. 7. Region of stability to the v reachable set under modified linear state variable feedback.

Fig. 8. Region of stability under game theoretic feedback control.

More analysis would be required to actually prove that $|x_1 + x_2| < 1 - \beta$ defines the largest guaranteed controllable set to the origin in the presence of noise. Note that for the "real" system, a state constraint of the form $|x_1| \leq (\alpha^2 M/r_m)(d - \ell)$ would be required to account for the finite gap. this further restricts the guaranteed controllable set in an obvious way.

For this system, both the limiting saturation approach and the game theoretic approach yield essentially identical results. This is not true in general. As will be reported elsewhere, using a simple nonlinear system, it can be shown that the game approach results in complete controllability for the system, whereas the limiting saturation approach results in a very restricted partial controllability.

ACKNOWLEDGMENT

The author is indebted to Dr. B. S. Goh and Professor George Leitmann for their suggestions in reference to the Liapunov stability condition.

REFERENCES

Brogan, W.L. (1964). Modern Control Theory. Quantum, New York.

Earnshaw, S. (1842). On the nature of molecular forces. Trans. of the Cambridge philos. Soc., 7, 97-112.

Grantham, W.J, and T.L. Vincent (1975). A controllability minimum principle. J. of Optimiz. Theory & Appl., 17, 93-114.

Gutman, S. and G. Leitmann (1976). Stabilizing feedback control for dynamical systems with bounded uncertainty. Proc. IEEE Conf. on Decision and Control.

Isaacs, R. (1965). Differential Games. Wiley, New York.

LaSalle, J., and S. Lefschetz (1961). Stability by Liapunov's direct methods. Academic Press, New York.

Leitmann, G. (1979). Guaranteed asymptotic stability for some linear systems with bounded uncertainties. To Appear J. Dynamical Systems, Measurement, and Control.

Ragade, R. K., and I.G. Sarma (1967). A game theoretic approach to optimal control in the presence of uncertainty. IEEE Trans. on Automatic Control, AC-12.

Sabnis, A. V., J.B. Dendy, and F.M. Schmitt (1975). A magnetically suspended large momentum wheel. J. of Spacecraft and Rockets, 12, 420-427.

NEW LINEAR PROGRAMMING METHODS AND THEIR APPLICATION TO OPTIMAL CONTROL

R. Gabasov* and F. M. Kirillova**

**Byelorussian State University, Minsk, USSR*
***Institute of Mathematics, Academy of Sciences, Minsk, USSR*

Abstract. The results of the authors and their colleagues on investigation of linear programming problems and their application are given in the report. The adaptive method for solving the general linear programming problem is described. The results of the numerical comparative experiment with the simplex method are presented. New methods for solving the large linear programming problems are given. The method for solving the typical linear optimal control problem is grounded.

Keywords. Optimal control, linear programming, numerical methods.

INTRODUCTION

In connection with the elaboration of numerical algorithms of optimal control, the authors, since 1970, have studied the analysis of the existing methods of linear and non-linear programming [1,2]. While at work, they have given the classification of linear programming methods and a number of new algorithms has been suggested for solving the general linear programming problem [3]. Lately, these algorithms have been modified for more precise account of the specificity of the problems under consideration [4-8]. A new method is called adaptive; as in each of its iterations the feasible solution and support change is closely connected with the information available at the given moment. The adaptive method is assumed as the basis for numerical methods of constructing optimal controls. The specificity of problems, consisting in a great number of variables possessed by each problem or in the phase constraints - if the latter are many, is accounted for in the modifications on optimal control.

The authors have elaborated the new decomposition methods to obtain the effective results of the realization of the large problems solution by a computer.

ADAPTIVE METHOD

The supporting method has been elaborated to solve linear programming problems in the canonic form [3]. Its peculiarity is in providing the opportunity to take into account a priori information on feasible solutions (direct and dual) and to stop the process of solving a problem with the help of suboptimality criterion on the feasible solution from some given neighborhood of the optimal solution. In [4], the supporting method with the adaptive normalization is given, which is also predetermined for the canonic form problems, but to differentiate between the simplex and the supporting methods it utilizes a more natural normalization of the admissible directions, depending on the current feasible solution. The support change extremum rule was deduced in this method for the first time. The rule is based on the supplementary information present on the iterations. The ideas mentioned above are developed in the present paper. The mathematical model and ensuring of the close connection with the whole collection of primary and current information and operations for their transformation are paid special attention to.

MATHEMATICAL MODEL THE BASIC NOTIONS

Let us consider the general linear programming problem in its natural form[1]

$$c'x \to \max, \; b_* \le Ax \le b^*, \; d_* \le x \le d^* . \quad (1)$$

Here c, x, d_*, d^* are n-vectors; b_*, b^* are m-vectors; A is $m \times n$-matrix, it is assumed that each vector is recorded in the form of a column; transposition operator (stroke) is used for obtaining linevector; symbols in vector forms mean the set of component inequalities.

The general problem of linear programming is also known in two forms:

normal

a) $c'x \to \max, \; Ax \le b, \; x \ge 0,$

and canonic (standard)

b) $c'x \to \max, \; Ax = b, \; x \ge 0.$

---

1. Problem (I) is known also as the interval linear programming problem [9].

a), b) models are obtained from (1) at the particular values of its parameters. On the other hand (I) can be reduced to each of the models, if (I) is expanded at the cost of the increase in number of variables and constraints.

The normal form of the general problem is suitable for the linear programming theory, for the duality theory in particular. The canonic form is often used to illustrate the theoretical aspects of many computing methods, in the simplex method for instance. Form (I) is more natural because it is particularly this form that most often occurs at modelling of the applied problems, at utilization of linear programming methods in nonlinear programming and at the attempt of the effective utilization of a priori information. So, we'll later consider on the methods of solving problem (I). The notion of feasible solution of such n-vector x, on which the main ($b_* \leq Ax \leq b^*$) and direct ($d_* \leq x \leq d^*$) restrictions are fulfilled is connected with the model of problem (I). The feasible solution x^ϵ is called ϵ-suboptimal solution, if $c'x^o - c'x^\epsilon \leq \epsilon$ where x^o is optimal solution of problem (I): $c'x^o = \max c'x$, the maximum is calculated on all the feasible solutions.

One of the main elements of the further suggested approach is the supposition that at solving the applied problems not only the mathematical model (I) is used but also a priori information concerning feasible solutions. This information reflects the real systems functioning experience, knowledge of specialists, guess-work and intuition of experts, the results of solution of simplified problems and so on. It exists in any applied situation, forming the necessary element of progress, natural development of knowledge and experience.

The supposition about an access to the initial information on feasible solutions is closely connected with model (I) being chosen. In practice, the result of each experiment is, as it is known, three numbers: pessimistic, most probable and optimistic estimates. Thus, vectors d_*, x, d^* are considered as pessimistic, most probable and optimistic estimates of the optimal solution. The vectors b_*, Ax, b^* are analogously considered in resource terms, indispensable for realization of the optimal solution. Therefore, even in cases, when the initial model has a) or b) form, accounting for initial information brings one to model (I).

The basis of the adaptive method[2] is the notion of a support.

<u>Definition.</u> Non-singular matrix $A_{sup} = A(I_{sup}, \mathcal{J}_{sup})$, composed of a_{ij}, $i \in I_{sup}$, $j \in \mathcal{J}_{sup}$ elements of matrix $A = A(I, \mathcal{J})$, $I = \{1, 2, \ldots, m\}$, $\mathcal{J} = \{1, 2, \ldots, n\}$ is called the problem (I) support.

This new notion has been derived from the analysis of the simplex method and it is the analogue of the notion basis. However, the support in general case, is a self-contained element of a method which does not depend on the feasible solution.[3] On the process of solving problem (I) the support changes along with the feasible solution. The composition of the initial support is included into a priori information on the support along with the feasible solution quality as determined from suboptimal estimate (which value is closely connected with the feasible solution and the support).

The effectiveness of each method for solving practical problem depends on three elements: 1) the degree of utilizing a priori information, 2) the degree of utilizing current information for its transformation, 3) correctly timed finish of the solution process after obtaining the pre-set characteristics. The first element of the adaptive method has been discussed above in accordance with the model type and the initial information on feasible solutions. The second and the third elements will be considered later on.

DIRECT ADAPTIVE METHOD

The following definition is the basis of this method: the pair $\{x, A_{sup}\}$ from the feasible solution and support of problem (I) is called the supporting solution.

The supporting solution is considered non-degenerated, if it is not degenerated along the basic $[b_*(I_H) < A(I_H, \mathcal{J})\, x(\mathcal{J}) < b^*(I_H)]$ and direct $[d_*(\mathcal{J}_{sup}) < x(\mathcal{J}_{sup}) < d^*(\mathcal{J}_{sup})]$ restrictions.

Here $I_H = I/I_{sup}$, $b_*(I_H) = \{b_{*i}, i \in I_H\}$.

Let $\{x, A_{sup}\}$ be the initial supporting solution. Let us calculate the lower $\omega_* = \omega_*(x) = b_* - Ax$ and the upper $\omega^* = \omega^*(x) = b^* - Ax$ discrepancy vectors, potential vectors $u' = u'(I_{sup}) = C'_{sup} A^{-1}_{sup}$ and estimate vectors $\Delta' = \Delta'(\mathcal{J}) = u'A(I_{sup}, \mathcal{J}) - c'$.

At first the objective function increment formula is found for the transformation of the supporting solution

$$c'\Delta x = c'\tilde{x} - c'x$$

where $\tilde{x} = x + \Delta x$ is some solution of problem (I).

On determining estimate and potential vectors we obtain

$$\begin{aligned} c'\Delta x &= u'A(I_{sup}, \mathcal{J})\,\Delta x - \Delta'\Delta x \\ &= [u'A(I_{sup}, \mathcal{J}_H) - \Delta'(\mathcal{J}_H)]\,\Delta x(\mathcal{J}_H) \\ &\quad + u'A(I_{sup}, \mathcal{J}_{sup})\,\Delta x(\mathcal{J}_{sup}), \\ \mathcal{J}_H &= \mathcal{J}/\mathcal{J}_{sup} \end{aligned} \tag{2}$$

2. O. S. Kostjukova benefitted the elaboration of the method much.

3. In the simplex method, the basis, as it is known, determines the basic solution single-valuedly. The simplex method uses (only) the basic solution.

As $b_* \leq A(x + \Delta x) \leq b^*$, then $\omega_* \leq A\Delta x \leq \omega^*$. The supporting groups of the last inequalities is equivalent to the relations:

$$A(I_{sup}, \mathcal{J}_{sup})\Delta x(\mathcal{J}_{sup}) + A(I_{sup}, \mathcal{J}_H)\Delta x(\mathcal{J}_H) = \mathcal{L}(I_{sup}),\ \omega_*(I_{sup}) \leq \mathcal{L}(I_{sup}) \leq \omega^*(I_{sup}) \quad (3)$$

Then

$$\Delta x(\mathcal{J}_{sup}) = A_{sup}^{-1}\mathcal{L}(I_{sup}) - A_{sup}^{-1}A(I_{sup}, \mathcal{J}_H)\Delta x(\mathcal{J}_H) \quad (4)$$

From (2) we have the following:

$$C'\Delta x = -\Delta'(\mathcal{J}_H)\Delta x(\mathcal{J}_H) + u'\mathcal{L}(I_{sup}) \quad (5)$$

Let a_i be i - matrix line of A, $a_i = A_*(i, \mathcal{J})_*$, $I_* = I_*(x) = \{i\in I : a_i'x = b_{*i}\}$, $I^* = I^*(x) = \{i\in I : a_i'x = b_i^*\}$, $I^\sim = I/(I_* \cup I^*)$; $\mathcal{J}_* = \mathcal{J}_*(x) = \{j\in\mathcal{J} : x_j = d_{*j}\}$, $\mathcal{J}^* = \mathcal{J}(x) = \{j\in\mathcal{J} : x_j = d_j^*\}$, $\mathcal{J}^\sim = \mathcal{J}/(\mathcal{J}_* \cup \mathcal{J}^*)$; $I_{*sup} = I_* \cap I_{sup}$; $\mathcal{J}_{*H} = \mathcal{J}_* \cap \mathcal{J}_H$

Optimality criterion. The relations

$$u(I_{*sup}) \leq 0,\ u(I^*_{sup}) \geq 0,\ u(I^\sim_{sup}) = 0;$$
$$\Delta(\mathcal{J}_{*H}) \geq 0,\ \Delta(\mathcal{J}^*_H) \leq 0,\ \Delta(\mathcal{J}_H) = 0 \quad (6)$$

are sufficient for the optimality of the supporting solution $\{x, A_{sup}\}$. In case of non-degeneratedness they are also the necessary condition.

Proof. Sufficiency. Vector $\mathcal{L}(I_{*sup})$ is not negative on set I_{*sup} and not positive on set I^*_{sup}. Consequently, inequality $u'\mathcal{L}(I_{sup}) \leq 0$ for all $\mathcal{L}(I_{sup})$ from (3) is true in conditions (6).

The vector $\Delta x(\mathcal{J}_{*H})$ - on the set $\mathcal{J}_{*H}$ is not negative and not positive on the set $\mathcal{J}^*_H$ as $\Delta x(\mathcal{J}_{*H}) = \tilde{x}(\mathcal{J}_{*H}) - x(\mathcal{J}_{*H}) \geq d_*(\mathcal{J}_{*H}) - d_*(\mathcal{J}_{*H}) = 0$; $\Delta x(\mathcal{J}^*_H) \leq d^*(\mathcal{J}^*_H) - d^*(\mathcal{J}^*_H) = 0$

Thus, $\Delta'(\mathcal{J}_H)\Delta x(\mathcal{J}_H) \leq 0$ for all admissible $\Delta x(\mathcal{J}_H)$.

Therefore, any admissible increment Δx does not result in increase of objective function value according to (5) $c'\Delta x \leq 0$. The supporting solution $\{x, A_{sup}\}$ is optimal.

Necessity. Let $\{x, A_{sup}\}$ be the non-degenerated optimal solution. Assuming that the relations (6) are not realized for the potential group, then at $\Delta x_H = 0$ a sufficiently small admissible variation of the vector $\mathcal{L}(I_{sup})$ occurs, that does not disturb the direct restrictions on support variables and the main non-support restrictions and such that $u'\mathcal{L}(I_{sup}) > 0$.

Analogously, the disturbance of relations (6) on the estimate group provides if $\mathcal{L} = 0$ the opportunity for constructing of the admissible variation $\Delta x(\mathcal{J}_H)$ at which $-\Delta'(\mathcal{J}_H)\Delta x(\mathcal{J}_H) > 0$. As the obtained inequalities contradict the supposition on optimality of solution x, then the optimality criterion is proved.

The maximum of function (5), calculated without considering any part of restructions on Δx, is the estimate of the solution deviation from the optimal solution. Let us calculate the maximum of function (5) with regard to the restrictions

$$d_*(\mathcal{J}_H) - x(\mathcal{J}_H) \leq \Delta x(\mathcal{J}_H) \leq d^*(\mathcal{J}_H) - x(\mathcal{J}_H),$$
$$\omega_*(I_{sup}) \leq \mathcal{L}(I_{sup}) \leq \omega^*(I_{sup}) \quad (7)$$

It is attained at

$$\Delta x_j = d_{*j} - x_j,\ \text{if } \Delta_j > 0;$$
$$\Delta x_j = d_j^* - x_j,\ \text{if } \Delta_j < 0;$$
$$\Delta x_j = 0,\ \text{if } \Delta_j = 0, j\in\mathcal{J}_H;$$
$$\mathcal{L}_i = \omega_{*i},\ \text{if } u_i < 0;$$
$$\mathcal{L}_i = \omega_i^*,\ \text{if } u_i > 0;$$
$$\mathcal{L}_i = 0,\ \text{if } u_i = 0, i\in I_{sup}, \quad (8)$$

and equals to the following value

$$\beta = \beta(x, A_{sup}) = \sum_{\Delta j>0,\, j\in\mathcal{J}_H} \Delta j(x_j - d_{*j}) + \sum_{\Delta j<0,\, j\in\mathcal{J}_H} \Delta j(x_j - d_j^*) + \sum_{u_i<0,\, i\in I_{sup}} u_i\omega_{*i} + \sum_{u_i>0,\, i\in I_{sup}} u_i\omega^*_i \quad (9)$$

which will be called suboptimality estimate of the supporting solution $\{x, A_{sup}\}$.

Suboptimality criterion. At $\beta \leq \epsilon$ the supporting solution $\{x, A_{sup}\}$ is ϵ-optimal solution of problem (I). For each x^{ϵ} of ϵ-optimal solution such support A_{sup} exists that suboptimality estimate β of the supporting solution $\{x^{\epsilon}, A_{sup}\}$ satisfies the inequality $\beta \leq \epsilon$.

The first part of this is proved above. The correctness of the second part can be deduced from the expansion of value β below.

Let $\beta > \epsilon$, where ϵ - is the precision of approximation to the optimal solution x^o according to the values of the objective function. The natural principle for improving of the supporting solution $\{x, A_{sup}\}$ which is assumed in the adaptive method, is the transition to the supporting solution $\{\bar{x}, \bar{A}_{sup}\}$ suboptimality estimate $\bar{\beta}$ of which is less than β.

Let us introduce a problem dual to (I):

$$b^* s - b_*' t - d_*' \upsilon + d^{*\prime} w \quad \min,$$

$$A'y - \upsilon + w = c, \; s - t - y = 0,$$

$$s \leq 0, \; t \geq 0, \; \upsilon \geq 0, \; w \geq 0 \quad (10)$$

It is not difficult to check that the vector $\{y, s, t, \upsilon, \omega\}$ with the components

$$y = y(I), \; s = s(I), \; t = t(I), \; \upsilon = \upsilon(\mathcal{J}),$$

$$\omega = \omega(\mathcal{J}): y(I_{sup}) = u; \; y(I_H) = 0;$$

$$s_i = y_i, \; t_i = 0, \text{ if } y_i \geq 0;$$

$$s_i = 0, \; t_i = -y_i, \text{ if } y_i < 0, \; i \in I;$$

$$\upsilon_j = \Delta_j, \; \omega_j = 0 \text{ if } \Delta_j \geq 0;$$

$$\upsilon_j = 0, \; \omega_j = -\Delta_j \text{ if } \Delta_j < 0, \; j \in \mathcal{J}, \quad (11)$$

is a dual feasible solution. As this vector is determined by support A_{sup} of problem (I) single-valuedly, it is called the dual feasible solution, accompanying the supporting solution $\{x, A_{sup}\}$ (accompanying dual feasible solution).

With the help of feasible solution x and the accompanying dual feasible solution (11) we obtain from (9): $\beta = (c'x^o - c'x) - (b^{*\prime}s^o - b_*' t^o - d_*' \upsilon^o + d^{*\prime} \omega^o - b^{*\prime}s + b_*' t + d_*' \upsilon - d^{*\prime} \omega)$.

Thus, suboptimality estimate β allows expansion

$$\beta = \beta_x + \beta_{sup}, \quad (12)$$

where $\beta_x = c'x^o - c'x$ is the non-optimality measure for the solution x, where $\beta_{sup} = b^{*\prime}s - b_*'t - d_*' \upsilon + d^{*\prime}\omega - b^{*\prime}s^o + b_*'t^o + d_*'\upsilon^o - d^*\omega^o$ is the non-optimality measure for the support A_{sup}.

Let the accompanying dual feasible solution for A^o_{sup} be the optimal basic dual solution of problem (I). We add the support A^o_{sup} to x. Then $\beta_{sup} = 0$, $\beta = \beta_x$ and $\beta = c'x^o - c'x$. By this, the necessary part of the suboptimal criteria given above is proved.

Due to the expansion (12) the components x, A_{sup} of the supporting solution can be improved independently from each other. A new feasible solution x is constructed in the form $\bar{x} = x + \theta \ell$, where ℓ is a suitable direction, θ is a maximum admissible step along ℓ. The direct methods differ by the principle of choice of a suitable direction. In the simplex method and in the direct supporting method from [3] a suitable direction is constructed from the derivative maximum principle along admissible directions, limited by a special (simplex) normalizing condition $\left(\sum_{j \in \mathcal{J}_H} |\ell_j| = 1, \ell_j \geq 0, j \in \mathcal{J}_H\right)$. It is possible to choose the normalizing condition for the other direct methods (for the method of projecting gradients, for instance). Nevertheless all these conditions are heuristic and are introduced into the method from outside and are neither closely connected with the model structure, nor with the situation occurring on the iterations. The normalization (7) is more natural. This normalization is constructed only according to the problem parametres and depends on the current supporting solution and ensures a more precise account of constraints.

In the general case solution (8) of the maximum increment problem (5) under constraints (7) will not be an admissible increment, but the vector $\ell = \{\ell(\mathcal{J}_{sup}), \ell(\mathcal{J}_H)\}$, constructed according to (8):

$$\ell_j = d_{*j} - x_j, \text{ if } \Delta_j > 0;$$

$$\ell_j = d_j^* - x_j, \text{ if } \Delta_j < 0;$$

$$\ell_j = 0 \text{ if } \Delta_j = 0, \; j \in \mathcal{J}_H;$$

$$\ell(\mathcal{J}_{sup}) = A^{-1}_{sup} \, \omega(I_{sup}) - A^{-1}_{sup} A(I_{sup}, \mathcal{J}_H)$$

$$\omega(I_{sup}) = \{\omega_i, i \in I_{sup}\}, \; \omega_i = \omega_{*i}$$

if $u_i < 0$;

$$\omega_i = \omega_i^* \text{ if } u_i > 0; \quad (13)$$

$$\omega_i = 0, \text{ if } u_i = 0, \; i \in I_{sup}$$

is a suitable direction for the feasible solution x.

In the maximizing problem with constraints the maximizing increment of the objective function accounts for the main aim and structure of the problem. It is more natural, than the maximization of the derivative, which being in essence a local characteristic of the function, is only indirectly connected with the initial aim of the problem.

The suitable direction (13) in case of the optimal support A_{sup} ($\beta_{sup} = 0$) and non-degenerate accompanying dual solution within one iteration results in the optimal solution[4] $\bar{x} = x + \ell$, as in this case the vector $\bar{x}$ coincides with the pseudosolution, constructed single-valuedly by the dual supporting method on to the optimal dual solution.

It is not known in the general case, whether the support is optimal. Therefore to construct $\bar{x}$ it is necessary to calculate θ step. On the motion along ℓ the restrictions which have not been accounted for can be disturbed while the maximum increment of the objective function (5) at the conditions (7) took place. The maximum admissible step along the direct support constraints is equal to $\theta_{jo} = \min \theta_j$, $j \in \mathcal{J}_{sup}$; $\theta_j = (d_{*j} - x_j)/\ell_j$, if $\ell_j < 0$; $\theta_j = (d^*_j - x_j)/\ell_j$, if $\ell_j > 0$; $\theta_j = \infty$, if $\ell_j = 0$. The maximum admissible step along the main non-support constraints is equal to $\theta_{i_o} = \min \theta_i$, $i \in I_H$; $\theta_i = \omega_{*i}/a'_i\ell$, if $a'_i\ell < 0$, $\theta_i = \omega^*_i/a'_i\ell$, if $a'_i\ell > 0$; $\theta_i = \infty$, if $a'_i\ell = 0$.

Thus, $\theta = \min \{1, {}_{i_o}, \theta_{j_o}\}$, where the unity means that constraints (7) should not be disturbed on the motion along ℓ.

The following relations $c'x^o - c'x = c'x^o - c'x - \theta c'\ell \le \beta - \theta c'\ell = (1 - \theta)\beta$ are performed on a new feasible solution $\bar{x} = x + \theta\ell$, i.e., at $(1 - \theta)\beta \le \epsilon$ the process of problem solution (I) ceases with the construction ϵ - optimal solution $\bar{x}$.

Let $(1 - \theta)\beta > \epsilon$. Both in the simplex method and in the direct supporting method [3] the support substitution was determined in the general case single-valuedly with the renewing of the feasible solution. If for the simplex method this operation is the only possible one due to the peculiarity of the basic solution, the other principles of the support substitution are acceptable for our method. Following the principle, accepted above, the support A_{sup} is substituted by the support A_{sup} to decrease the suboptimality estimate. It is clearly seen that a new operation is closely connected with the situation occurring on the iteration.

Let us choose some subsets I_o, $\mathcal{J}_o$, ($I_o \subset I_H$, $\mathcal{J}_o \subset \mathcal{J}_{sup}$). Among different supports $\bar{A}_{sup} = A(\bar{I}_{sup}, \mathcal{J}_{sup})$, such that $\bar{I}_{sup} \cap I_o = \phi$, $\mathcal{J}_{sup} \supset \mathcal{J}_o$ let us find the one, for which the suboptimality estimate of the supporting solution $\{\bar{x}, \bar{A}_{sup}\}$ is minimum. The demand, that all elements of the set $\mathcal{J}_o$ should be supporting ($\mathcal{J}_o \subset \mathcal{J}_{sup}$), and all elements of the set I_o should be non-supporting ($I_o \cap \bar{I}_{sup} = \phi$) is equivalent to the conditions

$$w(\mathcal{J}_o) = v(\mathcal{J}_o) = 0, \quad y(I_o) = s(I_o) = t(I_o) = 0 \tag{14}$$

Thus, the search of the support $\bar{A}_{sup}$ is reduced to the search of the optimal basic solution of the dual problem (10) with the registration of the additional conditions (14). Introducing the designations $I = I/I_o$, $\tilde{\mathcal{J}} = \mathcal{J}/\mathcal{J}$, $\tilde{y} = \{y(\tilde{I}), s(\tilde{I}), t(I), v(\bar{\mathcal{J}}), w(\tilde{\mathcal{J}})\}$, a new problem can be recorded in the form

$$f(\tilde{y}) = b^{*\prime}(\tilde{I})\, s(\tilde{I}) - b'_*(I)\, t(\tilde{I}) - d'_*(\tilde{\mathcal{J}})\, v(\tilde{\mathcal{J}})$$

$$+ d^{*\prime}(\tilde{\mathcal{J}})\, w(\tilde{\mathcal{J}}) \to \min, \quad A'(\tilde{I}, \mathcal{J}_o)\, y(\tilde{I}) = c(\mathcal{J}_o),$$

$$A'(\tilde{I}, \tilde{\mathcal{J}})\, y(I) + w(\tilde{\mathcal{J}}) - v(\tilde{\mathcal{J}}) = c(\tilde{\mathcal{J}})', \quad -y(\tilde{I}) + s(\tilde{I})$$

$$- t(\tilde{I}) = 0, \; s(\tilde{I}) \ge 0, \; t(\tilde{I}) \ge 0, \; v(\tilde{\mathcal{J}}) \ge 0,$$

$$w(\tilde{\mathcal{J}}) \ge 0 \tag{15}$$

On solving problem (15) the feasible solution $\mathcal{J}$ composed of the dual feasible solution components accompanying the support A_{sup} can be taken as an initial basic solution. The basis of the optimal basic solution $\tilde{y}^* = \{y^*(\tilde{I}), s(\tilde{I}), t^*(\tilde{I}), w^*(\mathcal{J})\}$ of problem (15) is the unknown support A_{sup}.

It is easy to prove that the estimate $\bar{\beta}_{sup}$ for a new support $\bar{A}_{sup}$ is equal to $\beta_{sup} = \beta_{sup} - [f(\tilde{y}) - f(\tilde{y}^*)]$. Thus the support $\bar{A}_{sup}$ is not worse than A_{sup} support as $f(\tilde{y}) \ge f(\tilde{y})$. Then $\bar{\beta} = (1 - \theta)\beta - [f(\tilde{y}) - f(y)]$.

It is natural to expect that the value $f(y) - f(y^*)$ will be bigger if the set $\mathcal{J}_{sup}/\mathcal{J}$ has the indexes j for which the number $x_j + \ell_j$ go out of the segment bounds $[d_{*j}, d^*_j]$ mostly (analogously for I_H/I_o):

$$\mathcal{J}_o = \{j, j \in \mathcal{J}_{sup}, \beta(x_j + \ell_j, [d_{*j}, d^*_j]) \le \xi\},$$

$$I_o = \{i, i \in I_H, \beta(a'_i(x + \ell), [b_{*i}, b^*_i]) \le \xi\} \tag{16}$$

Here $\rho(a, .)$ is a distance between sets; ξ is an iteration parameter.

Let us consider the problem dual to (15)

---

4. In this case the optimal solution problem (I) can be obtained on the optimal support A_{sup} and directly using the duality theory.

$$c'(\mathcal{J})\,x(\mathcal{J}) \quad \max,\ b_*(\tilde{I}) \le A(\tilde{I},\mathcal{J})\,x(\mathcal{J}) \le b^*(\tilde{I}),$$

$$d_*(\tilde{\mathcal{J}}) \le x(\tilde{\mathcal{J}}) \le d^*(\tilde{\mathcal{J}}) \quad (17)$$

which can be solved together with (15) starting with the supporting solution $\{\bar{x}, A_{sup}\}$. If the following constraints

$$d_{*j} \le x_j^* \le d_j^*,\ j \in \mathcal{J}_o,\ b_{*i} \le a_i' x^* \le b_i^*, \quad (18)$$

are observed on the optimal supporting solution $\{x^*, \bar{A}_{sup}\}$ of problem (17), then $\bar{A}_{sup}$ is the optimal support of problem (I) and x^* is the optimal solution. Consequently, such supporting components of solution x can be included into the set $\mathcal{J}_o$ that are for from the bound:

$$\mathcal{J}_o = \{j,\ j \in \mathcal{J}_{sup}\ ;\ d_{*j} + \xi < \bar{x}_j < d_j^* - \xi\} \quad (19)$$

or those, on which the direct constraints are the last to be disturbed on the motion along ℓ :

$$\mathcal{J}_o = \{j\ ;\ j \in \mathcal{J}_{sup},\ \theta j > \theta + \xi\} \quad (20)$$

Analogously to I_o :

$$I_o = \{i : i \in I_H,\ b_{*i} + \xi < a_i'\bar{x} < b_i^* - \xi\} \quad (21)$$

or

$$I_o = \{i : i \in I_H,\ \theta_i > \theta + \xi\} \quad (22)$$

On the other hand the choice of a suitable direction on the next iteration (for the supporting solution $\{x, A_{sup}\}$) depends on the choice of the support $\bar{A}_{sup}$. Let the conditions (18) not be fulfilled. Let the vector $x^* - \bar{x}$ be a suitable direction. If feasible cosolution $\bar{\Delta}$ accompanying the support $\bar{A}_{sup}$ is nondegenerated, then the vector coincides with the vector ℓ, constructed according to (13) for cosolution $\bar{\Delta}$. If feasible cosolution $\bar{\Delta}$ is degenerated, the vectors $x^* - \bar{x}$ and ℓ can fail to coincide, nevertheless in any case, both of them are optimal solution of the problem (5), (7), constructed on the support A_{sup}. It is clear from (17) that the maximum admissible step along the direction $x^* - \bar{x}$ is attained either on the direct constraints $j \in \mathcal{J}_o$, or on the main constraints $i \in I_o$. It also accounts for constructing $\mathcal{J}_o$ and I_o according to (19), (21). Hereby a non-zero step $\bar{\theta}$ occurs. At big ξ set $\mathcal{J}_o = I_o = \varphi$ and problem (17), (15) coincides with the initial one. At $\xi = 0$ and provided the step θ on the previous iteration is determined single-valuedly, we obtain $I_o = I_H - i_o$, $\mathcal{J}_o = \mathcal{J}_{sup}$, i.e., if $\theta = \theta_{i_o}$, $I_o = I_H$, $\mathcal{J}_o = \mathcal{J}_{sup} / j_o$ if $\theta = \theta_{j_o}$. These relations are true also for the case of the sets (20) - (22) but with an additional condition: $\theta_j < \infty$, $j \in \mathcal{J}_{sup}$, $\theta_i < \infty$, $i \in I_H$.

Let us consider the sets (16) from this point of view. At $\xi > 0$ it may occur, that $j_o \in \mathcal{J}_o$ for $\theta = \theta j_i$ and $i_o \in I_o$ for $\theta = \theta_{i_o}$. In this case, the step $\bar{\theta}$ can be equal to zero On the other hand, at sufficiently big ξ we have:

$\mathcal{J}_o = \mathcal{J}_{sup}$, $I_o = I_H$. Problem (15) has the only solution $\tilde{y}$, consequently, the support does not change. At $\xi = 0$ in the general case we have $|\mathcal{J}_o| = |\mathcal{J}_{sup}| - 1$, $|I_o| = |I_H| - 1$, even if the step θ on the previous iteration has been determined uniquely.

Note. It is not difficult to prove that the suboptimality estimate of a new supporting solution $\{x, A_{sup}\}$ is equal to $\bar{\beta} = c'x^* - c'\bar{x}$, where x^* is the optimal solution of problem (17).

Transition $\{x, A_{sup}\} \to \{\bar{x}, A_{sup}\}$ is called a big iteration of the method. It consists of small iterations of solution of the problem (15) [or of the problem (17)]. A big iteration is called regular if $\bar{\beta}_{sup} < \beta_{sup}$.

Theorem. The direct adaptive method is finite, if the number of non-regular big iterations of the problem is finite.

The proof of the statement is analogous to the one given in [3] for the case of the direct supporting method with the adaptive normalization.

In the general case, a priori information does not provide the opportunity for indicating the initial supporting solution. Three possibilities are discussed in [3,4] : 1) absence of any information, 2) the existence of the feasible solution without the support, 3) the existence of quasisolution.

NUMERICAL EXPERIMENT

In order to examine the principles put in the basis of the adaptive method, a numerical experiment was carried out. The direct adaptive method was compared to the simplex method. In the main experiment the problems $c'x \to \max$, $Ax \le b$, $d_* \le x \le d^*$, are considered. These problems have the following dimensions 20×30, 30×40, 40×50, 50×60, 60×70, 70×80, 80×90, 90×100, 90×200. The Parameters c_j, a_{ij}, b_i, d_{*j}, d_j^* of elements c, A, b, d_*, d^* of the problem were generated by the transducer of occasional numbers with the uniform distribution on the segments $-100 \leq c_j \leq 100$, $-100 \leq a_{ij} \leq 100$, $0 \leq b_i \leq 100$, $-100 \leq d_{*j} \leq 0$, $0 \leq d_j^* \leq 100$.

The solution process started each time from feasible solution $x = 0$. At the simplex normalization the starting feasible solution was compiled each time from the unit vector of conditions, corresponding to the free variables x_i^f $(Ax + x^f = b)$. In the adaptive one the starting support had dimensions 0×0.

After the parameters were generated the problem has been solved by the simplex method at first and then by the adaptive method. The optimal solution and the number of iterations were printed out in the simplex method. For the adaptive method the values of direct and dual objective functions and composition of support were printed out on each iteration. In more than one thousand instances passed through the computer, the number of iterations is in the average 2.8 times more in the simplex method than in the adaptive one.

The comparison on the number of iterations is important from the point of view of accumulation of round-off errors, caused by the transformation of supports. Nevertheless, the above adopted characteristic does not fully reflect the advantages of the adaptive method. In the experiment for the case 20×30 the dimension of supports in the adaptive method did not exceed 10×10, while in the simplex method the supports with dimensions 20×20 have always been used. Therefore, the first additional experiment on the calculation time was additionally carried out. The second additional experiment had the aim to study the effect of the number of variables (n) of the problem under the constant number of the basic constraints on the results. In the main experiment the fully filled matrices of the conditions A were generated. The third additional experiment was carried out, in which the sequence of problems with the given densities of A matrix filling was solved. At first one problem was solved by the simplex method and then - by the adaptive one for each filling. The aim of the fourth additional experiment was to study the effect of the problem size on the calculation results. The matrices were generated when fully filled. The fifth additional experiment was accomplished with the different values of the d bound of the direct constraints: $|x_j| \leq d$, $j = \overline{1,n}$.

In all the experiments the essential advantages of the adaptive method as compared to the simplex method are clearly seen.

LARGE PROBLEMS

Linear programming problem with a great number of variables. Modifications of the methods which have been introduced and which can be utilized in the large problems, are described below.

Utilizing a part of non-support variables for enlarging the iteration step. Let us consider the general problem of linear programming in its canonic form:

$$c'x \to \max, \ Ax = b, \ d_* \leq x \leq d^* \quad (\text{rank } A = m), \tag{23}$$

provided the size of the vector x considerably exceeds the size of the vector b, $n >> m$.

The main idea of the method being introduced is as follows: non-supporting variables with small estimates $|\Delta_j|$ should be utilized not for increasing the rate of changing the objective function along the suitable direction, but should be directed for assisting the supporting variables with the purpose of obtaining the maximum admissible step. Due to the minute change in the variables $|\Delta_j|$, the change in the rate of x_j of the objective function is also minute, but nevertheless, these changes considerably affect the rate of changing the support variables x_{ij}, $i \in \mathcal{J}_{sup}$, by which the value of the admissible step θ is determined. The feasible solution components with the big estimates $|\Delta_j|$ are left for forming the suitable direction according to the scheme of the adaptive method. It becomes possible to lessen the effect of round-off error on the process of the problem solution.

Let us choose the figure $\xi \geq 0$ - parameter and divide sets $\mathcal{J} = \{1,2,\ldots,n\}$, $\mathcal{J}_H$ into subsets $\mathcal{J}_M = \{j \in \mathcal{J}: |\Delta_j| \leq \xi\}$, $\mathcal{J}_{H\delta} = \{j \in \mathcal{J}_H: |\Delta_j| > \xi\}$, $\mathcal{J}_{HM} = \mathcal{J}_H \cap \mathcal{J}_M$. Let us construct the components of the admissible direction, corresponding to the big estimates:

$$\ell_j = d_{*j} - x_j, \text{ if } \Delta_j > \xi; \ \ell_j = d_j^* - x_j, \text{ if } \Delta_j < -\xi, \ j \in \mathcal{J}_{H\delta}. \tag{24}$$

On preserving the remaining non-supporting solution components, the objective function gains the maximum increment in the space of the variables x_j, $j \in \mathcal{J}_{H\delta}$.

The components ℓ_j, $j \in \mathcal{J}_M$ of the admissible direction ℓ are constructed to obtain the maximum step θ along ℓ:

$$\theta \to \max_{\ell_j, \, j \in \mathcal{J}_M}, \tag{25}$$

where

$$\theta = \min \{\theta_j, j \in \mathcal{J}_M; 1\},$$

$$\theta_j = \begin{cases} (d_{*j} - x_j)/\ell_j, \text{ if } \ell_j < 0, \\ (d_j^* - x_j)/\ell_j, \text{ if } \ell_j > 0, \\ \infty, \text{ if } \ell_j = 0, \ j \in \mathcal{J}_M, \end{cases} \tag{26}$$

and the variables ℓ_j, $j \in J_M$ satisfy the conditions of the admissible direction:

$$A_M \ell_M = B_\delta = 0, \quad B_\delta = A_{H\delta} \ell_{H\delta}. \qquad (27)$$

At $\theta = 1$ the feasible solution x is β - optimal, where

$$\beta = \sum_{\Delta_j > 0, j \in J_{HM}} (x_j + \ell_j - d_{*j}) + \sum_{\Delta_j < 0, j \in J_{HM}} \Delta_j (x_j + \ell_j - d_j^*) \le \xi \left(\sum_{\Delta_j > 0, j \in J_{HM}} (x_j + \ell_j - d_{*j}) + \sum_{\Delta_j < 0, j \in J_{HM}} (x_j + \ell_j - d_j^*) \right).$$

At a sufficiently small ξ the following inequality $\beta \le \epsilon$ is realized. If it happens that $\beta > \epsilon$, then ξ should be decreased and the problem solution (27) is proceeded at new ξ, z. At $\theta < 1$ a new feasible solution $\bar{x} = x + \theta \ell$ is constructed.

Let us discuss problem (25), (26) in more detail. The constraints (26), (27) for the step θ and the vector ℓ_M are obtained from the relations

$$d_{*M} - x_M \le \theta \ell_M \le d_M^* - x_M,$$

$$A_M \ell_M + b_\delta = 0, \quad 0 \le \theta \le 1.$$

If the last equity is multiplied on θ and new variables $S_M = \theta \ell_M$ are applied, then we obtain the following equivalent (25), (26) problem for calculating the maximum step:

$$\theta \to \max, \quad A_M S_M + \theta b \delta = 0, \qquad (28)$$

$$d_{*M} - x_M \le S_M \le d_M^* - x_M, \quad 0 \le \theta \le 1.$$

The following ensemble $\{\theta^*, S_M^*, A_{sup}\}$ can be assumed as the initial supporting problem solution (28), where A_{sup} is the support of the feasible solution x; $S_M^* = \{S^*_{sup}, S^*_{HM}\}$;

$S^*_{HM} = 0$;

$$S^*_{sup} = \theta^* A_{sup}^{-1} b_j;$$

$$\theta^* = \min \theta_j^*, \ j \in J_{sup},$$

$$\theta_j^* = (d_{*j} - x_j)/\ell_j^*, \text{ if } \ell_j^* < 0;$$

$$\theta_j^* = (d_j^* - x_j)/\ell_j^*, \text{ if } \ell_j^* > 0;$$

$$\theta_j^* = \infty, \text{ if } \ell_j^* = 0;$$

$$\ell_{sup} = -A_{sup}^{-1} b_\delta$$

The problem (28) is solved with the help of the supporting method with the adaptive normalization [4]. On the first iteration, the estimates of all non-supporting vector conditions α_j, $j \in J_{HM}$ of the problem (28) are equal to zero except the estimate $\Delta_\theta = -1$ of the vector $\mathcal{L}_\delta$. Consequently, at $\theta < 1$ only θ increases on the iteration among the non-supporting variables.

Let i_o be the index of the supporting variable s_{i_o}, which on the increase of θ on the segment [0, 1] was the first to appear on one of the borders of the direct constraints I_o is calculated according to the standard type formulae (26)]. Let us introduce the following designations:

$$J^* = [\, j \in J_M / J_{sup}:$$

$$s_j \ne d_{*j} - x_j, \text{ if } kx_{i_{o_j}} > 0;$$

$$s_j \ne d_j^* - xj, \text{ if } kx_{i_{o_j}} < 0\,],$$

$$k = \begin{cases} 1 \text{ if } s_{i_o} \text{ reaches the lower border,} \\ -1 \text{ if } s_{i_o} \text{ reaches the upper border,} \\ x(i_o, J_{HM}) = e_{i_o} A_{sup}^{-1} A_{HM} \end{cases}$$

If $J^* = \phi$, then the value of $\tilde{\theta}$, on which the increasing process of the non-supporting variable θ stopped, is equal to the maximum step along ℓ. The support $\tilde{A}_{sup}$, on which $J^* = \phi$ is added as an intermediate support to the feasible solution $\bar{x} = x + \theta \ell$, where $\ell = \{\ell_M, \ell_{H\delta}\}$, $\ell_M = \tilde{S}_M / \tilde{\theta}$, $\ell_{H\delta}$ – being calculated according to (24), if $\tilde{\theta} \ne 0$; $\ell = 0$, if $\tilde{\theta} = 0$; $\{\tilde{S}_M, \tilde{\theta}\}$ is the problem solution (28).

In the problem (23) we calculate the estimates $\tilde{\Delta}_j$ of the non-supporting vector conditions and the pseudosolution $\tilde{\chi}$, according to the support $\tilde{A}_{sup}$. If $\tilde{\chi}_{i_o} < d_{*i_o}$ at $k = 1$; $\tilde{\chi}_{i_o} > d^*_{i_o}$ at $k = -1$, then the deduction of the vector a_{i_o} from $\tilde{A}_{sup}$ according to the rules of the adaptive method results in improving the support A_{sup}. It can be stated, that in case of sign Δj = sign $\tilde{\Delta}_j$, $j \in J_{H\delta}$, the estimates $\tilde{\Delta}_j$, $j \in J_M$

do not affect the vector a_{j_o} choice, which will be into the support instead of a_{i_o} . Therefore, the adaptive method, on the support substitution, it is sufficient to calculate the dual step δ according to the estimates $\tilde{\Delta}_j$, $j \in \mathcal{J}_{H\delta}$. The iteration in the problem (23) is finished with the construction of the supporting solution $\{\bar{x}, A_{sup}\}$, $\bar{A}_{sup} = \left(A_{sup}/a_{i_o}\right) \cup a_{j_o}$.

If the pseudosolution $\tilde{\chi}$, constructed on the support $\tilde{A}$ satisfies the relations $\tilde{\chi}_{i_o} \geq d_{*i_o}$ at $k = 1$; $\tilde{\chi}_{i_o} \leq d^*_{i_o}$ at $k = -1$, then the immediate support, $\tilde{A}_{sup}$ is taken as the support $\bar{A}_{sup}$ for the feasible solution $\bar{x}$. It is possible to say that in this case, the problem (28) constructed according to the supporting solution $\{\bar{x}, \bar{A}_{sup}\}$ will have the solution with $c\theta > 0$ (it is assumed that $\bar{x}_{i_o}$ is the unique critical supporting variable in the supporting solution $\{\bar{x}, A_{sup}\}$).

In the case of $\mathcal{J}^* \neq \phi$ the special situation present before the iteration occurs after the iteration.

The utilization of a part of the non-supporting variables for increasing the iteration increment of the objective function. The main task of the iteration for all single-step optimization methods, in which the operations on the iteration are not coordinated with the operations on the subsequent iterations, is in the maximization of the increment of the objective function on the iteration. The components of an admissible direction, corresponding to the small estimates are used for solving this problem further on.

It can be derived from the physical sense of the estimate vector Δ that the increment of the objective function due to the change in the components of x_H is equal to $-\Delta'_H \ell_H = -\Delta'_{HM} \ell_{HM} -\Delta'_{H\delta} \ell_{H\delta}$. Let us construct the component $\ell_{H\delta}$ according to (24) . We assume that $\beta_{H\delta} = -\Delta'_{H\delta} \ell_{H\delta}$. The maximum on ℓ_{HM} for values $-\Delta'_H \ell_H$ is attained simultaneously with the maximum on $\theta \geq 0$, $S_{HM} = \theta \ell_{HM}$ for the value $-\theta \Delta'_H \ell_H = -\Delta'_{HM} S_{HM} + \theta\beta_{H\delta}$. The set of the admissible values for θ, S_M is described at the beginning of this section on Large Problems.

Thus, the maximum iteration increment of the objective function at some given direction $\ell_{H\delta}$ for the change in the variables $x_{H\delta}$ comes to the following problem

$$\theta\beta_{H\delta} - \Delta'_{HM} S_{HM} \rightarrow \max_{\theta, S_M} , \quad A_M S_M + \theta\beta g = 0,$$

$$d_{*M} - x_M \leq S_M \leq d^*_M - x_M . \qquad (29)$$

The problems (28), (29) are solved by the same method.

Component construction of the suitable direction. The set of the non-supporting indices $\mathcal{J}_H$ divided into several subsets $\mathcal{J}_1, \ldots, \mathcal{J}_k$ $\mathcal{J}_s \cap \mathcal{J}_p = \phi$, $s \neq p$, $\bigcup_{s=1}^{k} \mathcal{J}_s = \mathcal{J}_H$ for the non-degenerated supporting solution. To facilitate the calculations, let us consider the case of two sets $\mathcal{J}_1$, $\mathcal{J}_2$. As $\mathcal{J}_1$, $\mathcal{J}_2$ we can use, $\mathcal{J}_{H\delta}$, $\mathcal{J}_{HM}$, for instance.

Let us construct three vectors ℓ, ℓ_1, ℓ_2 . The vector ℓ is a suitable direction for the supporting solution $\{x, A_{sup}\}$, constructed according to the adaptive method:

$$\ell = \{\ell_{sup}, \ell_H\} , \quad \ell_{sup} = A^{-1}_{sup} A_H \ell_H ,$$

$$\ell_j = \begin{cases} d_{*j} - x_j, & \text{if } \Delta_j > 0, \\ d^*_j - x_j , & \text{if } \Delta_j < 0, \\ 0 , & \text{if } \Delta_j = 0, \ j \in \mathcal{J}_H \end{cases} \qquad (30)$$

In $\mathcal{J}^*_{sup} = \mathcal{J}_{sup} \cup \mathcal{J}_{*sup}$ we designate the indices set of the pseudosolution $\chi = x + \ell$ supporting components, which values are outside the borders of the direct constraints

$$\mathcal{J}^*_{sup} = \{j \in \mathcal{J}_{sup}, \chi_j > d^*_j\} ,$$

$$\mathcal{J}^*_{sup} = \{j \in \mathcal{J}_{sup}, \chi_j < d_{*j}\}$$

If $\mathcal{J}^{\chi}_{sup} = \phi$, then it is easy to show, that $x+\ell$ is the optimal solution of the problem (23) . We assume that the components of the vector $\ell_1 = \{\ell_{1\,sup}, \ell_{11}, \ell_{12}\}$ are equal to $\ell_{1sup} = \ell_1(\mathcal{J}_{sup}) = -A^{-1}_{sup} A_H \ell_2(\mathcal{J}_H)$, $\ell_{21} = \ell_2(\mathcal{J}_1) = 0$, $\ell_{22} = \ell_2(\mathcal{J}_2) = \ell(\mathcal{J}_2)$. The vector $\ell_2 = \{\ell_{2sup}, \ell_{21}, \ell_{22}\}$: $\ell_{2sup} = \ell_2(\mathcal{J}_{sup}) = -A^{-1}_{sup} A_H \ell_2(\mathcal{J}_H)$, $\ell_{21} = \ell_2(\mathcal{J}_1) = 0$, $\ell_{22} = \ell_2(\mathcal{J}_2) = \ell(\mathcal{J}_2)$, is constructed analogously.

The improvement of the supporting solution $\{x, A_{sup}\}$ is started with improving the feasible solution x along the suitable direction ℓ_1 : $x^1 = x + \theta^{(1)}_1 \ell_1$, where $\theta^{(1)}_1$ is the maximum admissible step, corresponding to the direction ℓ_1 for x. If $\theta^{(1)}_1 = 1$ and for all the components $x^1_j = x_j$, $j \in \mathcal{J}_2$, the optimality conditions are observed, then the feasible solution x^1 is the optimal solution of the problem (23) . Let us assume that these conditions are not observed.

Let ℓ_1 be the index of the critical supporting solution component, on which the step $\theta_1^{(1)}$ is attained. Two cases are possible: a) for $i = i_1$ and $\bar{x} = x^1$ the following conditions are observed.

$$i \in J^*_{sup} \quad \text{if } \bar{x}_i = d_{*i} \; ; \; i \in J^*_{sup} \quad \text{if } x_i = d^*_i \; ; \tag{31}$$

b) for $i = i_1$ and $\bar{x} = x^1$ the conditions (29) are not observed.

In the first case, the support A_{sup} is improved by removing vector a_{i_o} from A_{sup} and introducing the vector a_{j_o} instead of it. The index for a_{j_o} is calculated according to the supporting method with the adaptive normalization. The iteration ceases with the construction of a new feasible solution and a new support.

In case b) we proceed with improving the feasible solution x . The vector ℓ_2 is the suitable direction for the solution x^1. Applying ℓ_2 , we construct a new feasible solution

$$x^2 = x^1 + \theta_2^{(2)} \ell_2$$
$$= x + \theta_1^{(1)} \ell_1 + \theta_2^{(2)} \ell_2 \; , \tag{32}$$

where $\theta_2^{(2)}$ is the maximum admissible step from x along ℓ_2 . Let i_2 be the index of the critical supporting variable x^2 , which determined the step $\theta_1^{(2)}$. If for $i = i_2$ and $x = x^2$ the conditions (31) are observed, then, an improved support $\bar{A}_{sup} = (A_{sup}/a_{i_2}) \cup a_{j_o}$ is constructed for the feasible solution x^2 where the vector a_{j_o} is found according to the support method with the adaptive normalization.

If for $i = i_2$ and $x = x^2$ the conditions (31) are not observed, then it can be said, that $\theta_1^{(1)} < \theta_2^{(2)}$, $\theta_1^{(1)} < \theta$, where θ is the maximum admissible step from the feasible solution x along the direction ℓ .

Let the conditions (31) not be observed for $i = i_2$ and $\bar{x} = x^2$. We proceed with improving the feasible solution x . The direction ℓ_1 is suitable for the feasible solution x^2 . Therefore we construct the feasible solution

$$x^3 = x^2 + \Delta\theta_1^{(3)} \ell_1 = x + \theta_1^{(3)} \ell_1 + \theta_2^{(2)} \ell_2$$

where $\theta_1^{(3)} = \theta_1^{(1)} + \Delta\theta_1^{(3)}$, $\Delta\theta_1$ is the maximum admissible step along ℓ_1 , for the feasible solution x^2 . Let j_1 be the index of the supporting solution x^3 component, which determined the step $\Delta\theta_1$. If for $i = j_1$ and $\bar{x} = x^3$ the conditions (31) are observed, then we turn to an improved support $\bar{A}_{sup} = (A_{sup}/j_1) \cup j_1$. If for $i = j_1$ the conditions (31) are not observed, we can say, that $\theta_1^{(3)} > \theta_2^{(2)}$, $\theta_2^{(2)} < \theta$ and the vector ℓ_2 is the suitable direction for the feasible solution x^3 . We construct a new feasible solution x^4 and repeat the same arguments, described for the feasible solution x^2 [see (32)].

It can be proved that in finite number of steps the case a) obligatory occurs.

A big iteration ceases with the construction of a new supporting solution $\{\bar{x}, \bar{A}_{sup}\}$, where $\bar{x} = x^{(S)}$.

The rule for choosing a leading column a_{j_o} does not guarantee the non-degeneracy of a new supporting solution. Nevertheless, as in the direct supporting method with the adaptive normalization [4] , it can be stated that the existence of one critical variable $\bar{x}_{j_o}$ in the supporting solution $\{\bar{x}, \bar{A}_{sup}\}$ does not interfere with the construction of the following feasible solution on the next iteration $\bar{x}^{(k)} = \bar{x} + \theta_1^{(k)} \ell_1 + \theta_2^{(k-1)} \ell_2$ $\left(\text{or } \bar{x} + \theta_1^{(k-1)} \ell_1 + \theta_2^{(k)} \ell_2\right)$. At it $\theta_1^{(k)} + \theta_2^{(k-1)} > 0$ and the conditions (31) are observed for the critical supporting variable $\bar{x}_{i*}^{(k)}$.

OPTIMIZATION OF CONTROL

Optimal control methods have gained one of the central places in the scientific and technical literature along with the technology since twenty-five years. Nowadays a number of numerical methods for constructing the optimal conditions is known, but the problem of creating effective algorithms in this field is far from being solved yet. The modifications of the adaptive method (see Large Problems) applied to the linear optimal control problems are described below.

TERMINAL CONTROL PROBLEM

The adaptive method (see Large Problems) is applied for the optimization of the linear system in the class of pulse signals. Simultaneously, Pontrjagin's maximum principle and its generalization for ϵ – optimal controls is proved.

Problem Statement. Let us consider the control system, whose behavior within the time interval $T_1 = [0, \tau^*]$ can be described by the following equation

$$\frac{dz}{d\tau} = Qz + qv \; , \quad z(0) = x_o \; , \tag{33}$$

Here $z = z(\tau)$ is the state vector at the moment τ , $v = v(\tau)$ is the control value at the moment τ , Q , q are constants $n \times n$-matrix, and n-vector x_o is the initial state of the system.

Let us choose the set of time moments

$$\{\tau_t, t\epsilon T^*\}, \; t_o = 0, \; \tau_t > \tau_{t-1}, \; \tau_{t_1} = \tau_*$$

$$T^* = \{0, 1, \dots, t_1\}$$

on T_1 during which the observation after the system (33) state takes place. The set of available control $\upsilon(\tau)$ values at $\tau_t \leq \tau \leq \tau_{t+1}$ is designated via

$$u_t = \{u: f_* (t) \leq u \leq f^*(t)\} \quad (34)$$

The function $\upsilon(\tau)$, $\tau\epsilon T_1$ is called an available control, if it acquires the constant values $u(t)\,\epsilon U_t$ on each set $[\tau_t, \tau_{t+1}]$. An available control is called admissible if the trajectory $z(\tau)$, $\tau\epsilon T_1$ of the system (33), corresponding to it, is true for the equality

$$Hz(\tau_*) = g, \quad (35)$$

where H is the given $m \times n$ matrix and g is the given m-vector, rank $H = m$.

The quality of the admissible control $\upsilon(\tau)$, $\tau\epsilon T_1/\tau_*$, can be estimated with the help of the terminal functional

$$\mathcal{J}(\upsilon) = c'z(\tau_*) \quad (36)$$

determined via the objective function on the terminal states of the system (33). The admissible control $\upsilon^o(\tau)$, $\tau\epsilon T_1/\tau_*$, providing the maximum for the quality criterion (36) among all admissible controls is called optimal:

$$\mathcal{J}(\upsilon^o) = \max \mathcal{J}(\upsilon) \quad (37)$$

The trajectory $z^o(\tau)$, $\tau\epsilon T_1$, engendered by the optimal control, is called the optimal trajectory of the system (33).

Our problem is to construct the algorithm for calculating the optimal controls for the problem (33-37).

Taking into consideration the utilization of the linear programming methods, let us turn to another equivalent presentation of the problem, being studied. We shall use the known presentation of the system (33) solutions

$$z(\tau) = F(\tau) F^{-1}(s) z(s) + \int_s^\tau F(\tau) F^{-1}(\sigma) q \upsilon(\sigma) d\sigma,$$

$$\frac{dF}{d\tau} = QF, \; F(0) = E \quad (38)$$

(E is the unit diagonal $n \times n$ matrix)

From (38) we obtain for the observation of moments τ_t as follows

$$z(\tau_{t+1}) = F(\tau_{t+1}) F^{-1}(\tau_t) z(\tau_t) + \int_{\tau_t}^{\tau_{t+1}} F(\tau_{t+1}) F(\sigma) q \upsilon(\sigma) d\sigma. \quad (39)$$

Remembering that $\upsilon(\tau) \equiv u(t)$, $\tau_t \leq \tau \leq \tau_{t+1}$, and introducing the designations $x(t) = z(\tau_t)$, $A = \bar{F}(\tau_{t+1}) \bar{F}^{-1}(\tau_t)$,

$$b = \int_{\tau_t}^{\tau_{t+1}} F(\tau_{t+1}) F^{-1}(\sigma) q \, d\sigma$$

from (39) we get a recurrent equation

$$x(t+1) = Ax(t) + bu(t),$$

$$t\,T = T^*/t_1,$$

$$x(0) = x_o \quad (40)$$

If the control system states only at the observation moments τ_t, $t\epsilon T^*$ are of interest, and the control classes, introduced above, are applied, then (40) is equivalent to (33).

Thus, the problem (33) – (37) acquires the following form

$$c'x(t_1) \rightarrow \max, \quad x(t+1) = Ax(t) + bu(t),$$

$$x(0) = x_o, \quad f_*(t) \leq u(t) \leq f^*(t), \quad t\epsilon T,$$

$$Hx(t_1) = g \quad (41)$$

If the formula (38) is used, we can turn to the general linear programming problem in the canonic form relative to the unknown values $u(t)$, $t\epsilon T$. Without stating the details of reformulation, we shall present the result, effluent from the Large Problems section.

Problem solution. The set of moments $T_{sup} \subset T = \{0, 1, \dots, t_1 - 1\}$ is called the problem (41) support, if the system

$$x(t+1) = Ax(t) + bu(t), \quad x(0) = 0,$$

$$Hx(t_1) = 0, \quad t\epsilon T \quad (42)$$

possesses at $u(t) \equiv 0$, $t\epsilon T_H = T/T_{sup}$, only a trivial solution $u(t) \equiv 0$, $t\epsilon T_{sup}$, but at any $t^*\epsilon T_H$ the system (42) allows for $u(t) \equiv 0$, $t\epsilon T_H/t^*$ a non-trivial solution $u(t) \neq 0$, $t\epsilon T_{sup} \cup t^*$.

The set T_{sup} is the support then and only then, when

$$\det P \neq 0 \ , \quad P = \{HA^{t_1-t-1} b \ , \ t \in T_{sup}\} \qquad (43)$$

The pair $\{u(.) \ , \ T_{sup}\}$ from the admissible control and support is called the support control. The signals $u(t) \ , \ t \in T_{sup}$ will be called support ones; $u(t) \ , \ t \in T_H$ are non-support signals. The supporting control is called non-degenerated, if its support signals are non-critical

$$f_*(t) < u(t) < f^*(t) \ , \quad t \in T_{sup}$$

Along with the support control $\{u(.) \ , \ T_{sup}\}$ we shall discuss the admissible control $u^*(.)$ and calculate the objective function increment for the problem (41) :

$$\Delta \mathcal{J}(u) = \mathcal{J}(u^*) - \mathcal{J}(u) = - \sum_{S \in T_H} \Delta(s) \, \Delta u(s) \ ,$$

where $\Delta u(s) = u^*(s) - u(s)$ is the control increment, $-\Delta(s) = \psi'(s)\, b$ is the co-control, $\psi(s)$ (the cotrajectory) is the solution of the conjugated equation

$$\psi'(s-1) = \psi'(s) A \qquad (44)$$

with the initial condition

$$\psi'(t_1-1) = c' - \upsilon' H \ , \qquad (45)$$

υ is the m-vector of the potentials $\upsilon' = \left| c' A^{t_1-t-1} b \ , \ t \in T_{sup} \right| P^{-1}$

The inequality is true.

$$\mathcal{J}(u^\circ) - \mathcal{J}(u) \leq \beta, \qquad (46)$$

$$\beta = \sum_{\Delta(t)>0} \Delta(t) \left[-f_*(t) + u(t) \right] + \sum_{\Delta(t)<0} \Delta(t) \left[-f^*(t) + u(t) \right]$$

The admissible control $u^\epsilon(.)$ is called ϵ-optimal (suboptimal), if $c'x^0(t_1) - c'x^\epsilon(t_1) \leq \epsilon$, where $x^\epsilon(t) \ , \ t \in T^*$, is the system (41) trajectory, corresponding to the control $u^\epsilon(.) \ , \ u^0 = \{u^0(t) \ , \ t \in T\} \ , \ x^0(t) \ , \ t \in T^*$ are optimal control and the problem (41) trajectory.

It follows from (46) that

1) Suboptimality criterion. For ϵ-optimality of the admissible control $u(.)$ it is necessary and sufficient, that at some support T_{sup} the inequality $\beta \leq \epsilon$ is realized.

2) Optimality criterion. The relations

$$\Delta(t) = 0 \quad \text{at} \quad f_*(t) < u(t) < f^*(t) \ ;$$

$$\Delta(t) \geq 0 \quad \text{at} \quad u(t) = f_*(t) \ ;$$

$$\Delta(t) \leq 0 \quad \text{at} \quad u(t) = f^*(t) \ , \quad t \in T_H \ ,$$

are sufficient and in case of non-degeneracy even necessary for the optimality of the support control $\{u(.) \ , \ T_{sup}\}$.

The optimality criterion can be formulated in the terms of the maximum discreet principle: for the optimality of the support control $\{u(.) \ , \ T_{sup}\}$ it is sufficient that it should satisfy the maximum condition

$$\psi'(t)\, bu(t) = \max_{f_*(t) \leq u \leq f^*(t)} \psi'(t)\, bu \qquad (47)$$

where $\psi(t)$ is the cotrajectory, determined above; each non-degenerated optimal control $\{u^\circ(.) \ , \ T_{sup}\}$ satisfies the maximum condition (47).

The discrete maximum principle being introduced differs from the traditional forms in the definite indication of the cotrajectory, calculating which no extremal problems are solved. In connection with this, there are certain limitations of the maximum principle in the necessary part. The following form of the maximum principle lacks this shortcoming: for the optimality of the admissible control $u(t) \ , \ t \in T$, it is necessary and sufficient for the support T_{sup} to exist, then the support control $\{u(.) \ , \ T_{sup}\}$ satisfies the maximum (47) condition.

The maximum principle (47) is obtained from the optimality criterion of the support control. Basing on the suboptimality criterion, let us formulate the condition (47) analogue for the case of ϵ-optimal control.

For ϵ-optimality of the admissible control $u(t) \ , \ t \in T$, it is necessary and sufficient that at some support T_{sup} , such a function

$$\epsilon(t) \geq 0 \ , \quad t \in T \ , \quad \sum_{t \in T} \epsilon(t) \leq \epsilon$$

should exist, that

$$\psi'(t)\, bu(t) = \max_{f_*(t) \leq u \leq f^*(t)} \psi'(t)\, bu - \epsilon(t) \ , \quad t \in T \ ,$$

where $\psi(t)$ is the solution of the problem (44), (45) .

In the theory of optimal control, the conditions for ϵ-optimality have not been studied before. The result, being obtained, provides some stability characteristic for the maximum principle (according to the change in the control quality, the type of excitements for the maximum principle within the optimal control region is indicated).

Let us turn to the main problem. We assume that the support control $\{u(.), T_{sup}\}$ does not satisfy the optimality criterion and the estimate β of its deflection from the optimal control exceeds the set precision of the approachment ϵ . In the discrete problems (41) , obtained from the continuous (33) - (47) with the help of the time quantization, the figure t_1 , at the precise approximations, considerably exceeds m . Therefore, for solving the problem, the method from the section on Large Problems is natural. Let us construct the sets according to the co-control $\Delta(t)$, $t\in T$ and some given value α .

$$T_M = \{t: |\Delta(t)| \le \alpha\},$$

$$T_{H\delta} = \{t: |\Delta(t)| > \alpha\}.$$

Let us discuss the problem:

$$-\theta \sum_{t\in T} \Delta(t)\Delta u(t) \to \max ;$$

$$x(t+1) = Ax(t) + b\Delta u(t) \;;\; t\in T$$

$$Hx(t_1) = 0 \;,\; x(0) = 0 \;,$$

$$f_*(t) - u(t) \le \theta\Delta u(t) \le f^*(t)\, u(t) \;,\; t\in T_M \;,$$

$$\begin{aligned} \Delta u(t) &= f^*(t) - u(t) \;,\; \text{if } \Delta(t) < -\alpha, \\ \Delta u(t) &= f_*(t) - u(t) \;,\; \text{if } \Delta(t) > \alpha, \end{aligned} \quad t\in T_{H\delta}$$

This problem is equivalent to the next linear programming problem[5]

$$-\sum_{t\in T_M} \Delta(t)\Delta u(t)$$

$$-\theta \sum_{t\in T_{H\delta}} \Delta(t)\Delta u(t) \to \max_{\theta,\, \Delta u(t),\, t\in T,}$$

$$\sum_{t\in T_M} HA^{t_1-t-1} b\Delta u(t) = -\theta\tilde{b} \;;$$

$$f_*(t) - u(t) \le \Delta u(t) \le f^*(t) - u(t) \;,\; t\in T_M, \quad (48)$$

where $\tilde{b} = \sum_{t\in T_{H\delta}} HA^{t_1-t-1} b\Delta u(t)$.

The support solution can be taken as the initial support solution

$$\{\Delta u^1(t),\, t\in T_M \,,\, T_{sup}\},$$

where

$$\Delta u^1(t) = \theta^1 \Delta\tilde{u}(t)$$

$$\Delta\tilde{u}(t) = \begin{cases} 0 \text{ if } \Delta(t) = 0, \\ f^*(t) - u(t) \text{ if } \Delta(t) < 0,\; t\in T_{HM} = T_M / T_{sup} \\ f_*(t) - u(t) \text{ if } \Delta(t) > 0, \end{cases}$$

$$\Delta\tilde{u}_{sup} = -P^{-1}\tilde{b} - P^{-1}\left[\sum_{t\in T_{HM}} HA^{t_1-t-1} b\Delta\tilde{u}(t)\right]:$$

$$\theta^1 = \min \theta(t) \,,\, t\in T_{sup} \,,$$

$$\theta(t) = \begin{cases} [f^*(t) - u(t)]/\Delta\tilde{u}(t) \text{ if } \Delta\tilde{u}(t) > 0, \\ [f_*(t) - u(t)]/\Delta\tilde{u}(t) \text{ if } \Delta\tilde{u}(t) < 0, \\ \infty \text{ if } \Delta\tilde{u}(t) = 0, \end{cases}$$

the matrix P is set by the formula (43) .

The problem (48) realizes the problem from the section on Large Problems. Let the solution of the problem (48) be ceased on the support solution

$\{\Delta\bar{u}(t),\, t\in T_M;\, \bar{\theta} \,,\, T_{sup}\}$. The control

$\bar{u} = u + \Delta\bar{u}$ where $\Delta\bar{u} = \{\Delta\bar{u}(t),\; t\in T_M;$

$\bar{\theta}\Delta u(t),\, t\in T_{H\delta}\}$ is admissible in the problem (41)

$$\Delta \mathcal{J}(u) = \mathcal{J}(\bar{u}) - \mathcal{J}(u) = -\sum_{t\in T_M} \Delta(t)\Delta\bar{u}(t)$$

$$-\bar{\theta} \sum_{t\in T_{H\delta}} \Delta(t)\Delta u(t) \ge 0$$

(in non-degenerated case, it is strictly more than zero). The following inequality $\mathcal{J}(u^\circ) - \mathcal{J}(u) \le \beta$ is true, where

$$\beta = \sum_{\substack{\bar{\Delta}(t)<0 \\ t\in T}} \bar{\Delta}(t)\, [\bar{u}(t) - f_*(t)]$$

$$+ \sum_{\Delta(t)<0,\, t\in T} \bar{\Delta}(t)\, [\bar{u}(t) - f^*(t)] \,,$$

$$\bar{\Delta}(s) = \psi'(s)\, b \;,\; \psi(s)$$

is the solution of the conjugated system (44) with the initial condition (45) where

$$\upsilon' = \left\{ c' A^{t_1-t-1} b \;,\; t\in\bar{T}_{sup} \right\} \bar{P}^{-1} \,,$$

$$\bar{P} = \left\{ HA^{t_1-t-1} b \;,\; t\in\bar{T}_{sup} \right. .$$

If $\beta \le \epsilon$ then the control $\bar{u}$ is ϵ-optimal. In the reverse case the problem solution is proceeded according to the scheme in Large Problems.

5. In this problem the vectors are readily generated according to the initial H, A, b .

The methods from the foregoing paragraphs have been utilized in constructing the algorithms for the linear optimal control problems with the phase constraints. The allowance of the phase and mixed constraints demands for some modifications in the adaptive method, in particular, for those connected with the application of the working support matrix with the dimensions not exceeding the given value.

REFERENCE

Dantzig, G. Linear programming, its application and generalization. M. "Progress," 1964.

Kantorovitch, L. V., Mathematical methods for arrangement and planning of production. L. LGU Publishing House. 1939.

Gabasov, R., Kirillova, F. M., Linear Programming Methods. P. I. General Problems. BGU Publishing House, 1977.

Gabasov, R., Kirillova, F. M., Linear Programming Methods. P. II. Transport Problems. BGU Publishing House, 1978.

Gabasov, R., Kirillova, F. M., Kostynkova, O. I., On the Methods for Solving the General Linear Programming Problem. Preprint N 14 (30). Institute of Mathematics, Minsk, 1977.

Gabasov, R., Kirillova, F. M., The Construction of the Suboptimal Transport Problem Solutions. Engineering Cybernetics, N 6 (1975), p. p. 38-42.

Gabasov, R., Kirillova, F. M., Kostynkova, O. I., Adaptive methods for solving large linear programming problems. Preprint N 11 (43). Institute of Mathematics., Minsk, 1978.

Gabasov, R., Kirillova, F. M., Kostynkova, O. I., The method for solving the general linear Programming problem, Dokl, of Academy of Sciences of the BSSR, XXII, N 3 (1979), p. p. 197-200.

Charnes, A., Granot, D., Granot, G. A Primal Algorithm for the Internal Linear Programming Problems. Linear Algebra and its Applications, 17, N 1 (1977).

OPTIMAL CONTROL SOLUTION OF THE AUTOMOTIVE EMISSION-CONSTRAINED MINIMUM FUEL PROBLEM

A. R. Dohner

Electronics Department, General Motors Research Laboratories, Warren, MI, USA

Abstract. The automotive industry is confronted with the conflicting goals of improving fuel economy, reducing exhaust emissions, and maintaining vehicle driveability. Difficulties arise in the application of optimal control theory to this problem because transient models of the fuel, emission, and driveability responses do not exist. To circumvent these difficulties, the mathematical models are replaced by a sophisticated experimental test setup. To demonstrate the applicability of the optimal control approach without a mathematical model, the problem of the hot-start optimization of fuel economy subject to emission constraints problem is solved. Air-fuel ratio and spark advance are employed as the controls. Operational considerations necessitate the direct incorporation of the feedback control functions into the gradient-type solution algorithm. For this case, the results show that the oxides of nitrogen (NO_x) are controlled primarily through air-fuel ratio scheduling. The solution of this problem demonstrates the feasibility of the experimental optimal control approach. The second problem involves the cold-start portion of the Federal Test Procedure (FTP). The transient influences of the engine and catalytic converter warmup are analyzed by the optimization procedure and are reflected in the optimal feedback functions. For this case, the control of hydrocarbons (HC) and carbon monoxide (CO) is accomplished by minimizing engine-out concentrations immediately after startup. As the system warms up, the controls are gradually adjusted to reduce the fuel rate and to take advantage of the oxidation in the exhaust manifold and catalytic converter. The NO_x control is accomplished primarily through lean air-fuel ratio scheduling during hot operation. Finally, the hot-start optimization program is generalized to include an explicit surge-type driveability constraint on the controls. Comparison of the results of the hot-start problems reveals the trade-off between fuel economy and driveability. The imposition of the driveability constraint results in emission control primarily by retarding the spark advance and less by lean air-fuel ratio scheduling. Consequently, the driveability constraint causes a slight fuel economy penalty.

Keywords. Optimal control; automobiles; internal combustion engines; iterative methods; optimization; system theory.

INTRODUCTION

Today the automobile industry is confronted with the conflicting goals of improving fuel economy and reducing exhaust emissions of hydrocarbons (HC), carbon monoxide (CO), and oxides of nitrogen (NO_x). As the federal emission standards become more stringent, improvements in fuel economy become more elusive. Furthermore, as the engine controls are adjusted for better fuel economy and lower emissions, the driveability of the vehicle often is a problem. Because of the conflicting nature of these three considerations, the optimal solution for a particular engine/vehicle configuration is far from obvious. This thesis presents a methodological approach based upon optimal control theory to determine the best fuel economy-emission-driveability compromise.

In recent years many designs have been incorporated into the engine hardware, and exhaust aftertreatment devices have been added to the vehicle to accomplish the goal of improved fuel economy subject to emission and driveability constraints. Each of these hardware configurations has an optimal fuel economy for a particular set of constraint values. The control problem involves the determination of the control systems and the feedback control functions (typically for air-fuel ratio [A/F], spark advance [SA], and exhaust gas recirculation [EGR]) to achieve this optimum.

It is well-known to emission engineers that the influences of the controls (A/F, SA, EGR) on the emissions and fuel economy vary with engine operating conditions, e.g., speed, load, temperature. Driveability is also sensitive to the control values. Typically the control settings which minimize fuel consumption and NOx production also result in poor driveability. Resolving the trade-offs between fuel economy, emissions and driveability is particularly difficult because an accurate mathematical model of the complete engine/converter/vehicle system does not exist.

In the past few years several theoretical treatments of the optimal automotive engine control problem have appeared in the literature (A1,B1,C1,P1,R2).[1] All of these approaches employ steady-state data to approximate the fuel economy and cumulative emissions over the Federal Test Procedure (FTP) driving schedule (a twenty-two minute vehicle speed reference indicative of typical city and highway conditions).

These approaches represent an important first step in the solution of this problem. However, none of these approaches considers the transient system behavior during the FTP, which limits their value by one or more of the following considerations. The first limitation is the cold-start problem. Up to 80% of the total HC and CO are emitted during the cold-start portion of the test. Since only steady-state data are employed, treatment of the cold-start must be excluded from the optimization. The second limitation is the treatment of the influences of exhaust aftertreatment on emissions. The conversion efficiency of catalytic converters varies with A/F and temperature, which is dependent upon speed, load, SA, and EGR. Converter warmup characteristics exert a strong influence on cold-start emissions. The third limitation is that the driveability resulting from use of the optimal control laws has been ignored for the most part. In many regimes of engine operation, the optimization pushes the controls into an area of marginal combustion stability where, although NOx and fuel consumption may be minimum, the driveability is unacceptable. The fourth limitation is that these procedures do not directly address the compromises in the feedback control functions required by their physical implementation. Finally, these procedures typically require detailed engine maps which are costly and time-consuming to develop.

Prabhakar, et.al., (P1) present one of the first theoretical treatments of the automotive engine control problem. Curve fitting techniques are employed to develop analytical representations of experimental steady-state engine-out emissions and fuel consumption data as functions of engine operating variables. These emissions and fuel consumption curves are then used with a mathematical vehicle model to determine optimal feedback control functions during the FTP.

Rishavy, et.al., (R2) consider a linear programming problem using a small set of engine speed-load points to approximate the test schedule. A simple pass-through-ratio catalytic converter simulation is used in conjunction with the steady-state engine map data. Auiler, et. al., (A1) present a dynamic programming problem which again approximates the test schedule with a small set of engine speedload points. The emission contributions at each speedload point are allocated to maximize the projected fuel economy over the test schedule. A power-train simulation is used to select the engine speed-load points. The corresponding engine calibration is then inferred from the optimal emission allocation. The data-base problem for this approach is considered by Baker and Daby (B1), who also present a more heuristic approach to optimal engine calibration. Their results agree to some extent with those presented by Auiler (A1).

Cassidy (C1) also employs a small number of engine speed-load points to approximate the test schedule. Lagrange multipliers are used to treat the emission constraints. An on-line parameter optimization procedure is used to solve the empirical emission-constrainted minimum fuel problem at each steady-state speed-load point. After the parameter optimization is performed at each speed-load point, the Lagrange multipliers are adjusted if the projected emissions do not satisfy the constraints.

The goal of an optimization procedure is the optimized engine vehicle system, and the performance must be optimized over the entire transient emission and fuel economy test schedule, not just at selected speed-load points. The cold-start portion of the test, including the dynamics of the exhaust aftertreatment device, must be accounted for by the optimization. Driveability must be imposed as a formal constraint on the system throughout the test. In fact, all significant transient phenomena must be addressed by the optimization. None of the previously published approaches are capable of this due to either their basic nature or the lack of the required complex transient model of the entire system.

THE OPTIMAL CONTROL PROBLEM

In the Introduction, the approaches of various investigators to the automotive engine control problem are summarized. None of these studies have addressed the transient aspects of the problem, i.e., the transient driving schedule, the cold-start portion of the test, the controller dynamics, etc.

[1]Numbers in parentheses denote listings in the Reference section.

Either the values of the control settings are determined at selected steady-state speed-load points (A1, C1, D2, R2), or the system is modeled from steady-state data, and then the optimal control problem is solved employing this model (P1). The deficiencies of both these approaches, which are enumerated in the Introduction, result from an inadequate representation of the transient system behavior.

In this section these limitations are circumvented. The automotive engine control problem is posed in the framework of modern optimal control theory assuming the system responses of interest can be measured continuously throughout the federal fuel economy and emissions test. This obviates the need for the nonexistent, complete mathematical system model of the transient behavior during the test. The cumulative emissions are taken to be the states, and the cumulative fuel consumption is taken to be the cost. It is assumed that the system equations depend only on the controls and time and not on the states (the cumulative emissions) themselves.

In the remainder of this section, the necessary conditions of optimality are stated and a gradient-type optimization algorithm is defined. As will be shown, this gradient-type algorithm utilizes the special structure of this problem formulation to eliminate unnecessary evaluations of the system equations. The system equations are evaluated experimentally and are, therefore, relatively expensive in terms of test time and test facilities.

Later in the test, this gradient procedure is applied to the automotive minimum fuel problem with emission and driveability constraints. However, first the federal fuel economy and emissions test must be specified in the context of the minimum fuel problem, and the necessary conditions of optimality must be developed.

The Federal Fuel Economy and Emission Test (FTP)

The calculation of emissions and fuel economy by the Environmental Protection Agency (EPA) is a somewhat involved procedure described in the Federal Register (C6). The test is summarized briefly in this section, and the equations are presented in a form compatible with the minimum fuel formulation. The federal fuel economy and emissions test consists of two parts. First, the urban portion of the test is run to determine the emissions and the urban fuel economy. Next the highway portion is run to determine the highway fuel economy; during this part of the test the emissions are ignored.

The driving schedule (vehicle speed-time profile) is prescribed throughout the test and consists of idles, accelerations, cruises, and decelerations of various durations. The urban driving schedule, known as the LA-4 (S2), consists of 23 driving cycles. In this context a driving cycle is defined as the period from the beginning of one idle (vehicle speed = 0 MPH) to the beginning of the next. Cycles 1-5 are called the cold-start portion, cycles 6-18 are called the stabilized portion, and cycles 19-23 are called the hot-start portion (and are identical to cycles 1-5).

The test is conducted as follows (C6). After the vehicle has "soaked" at room temperature for 12 hours, cycles 1-18 are driven on a chassis dynamometer. The engine is shut down for 10 minutes (the hot soak), and then cycles 19-23 are driven. After the urban portion of the test, the highway portion is driven. The grams per mile of the emission constituents are defined in the Federal Register as the weighted sums of the grams of emissions from the cold-start, hot-start, and stabilized portions divided by the distance driven:

$$HC(g/mi)= \left[\frac{0.43*HC(C)+1.0*HC(S)+0.57*HC(H)}{7.5}\right] \quad (1)$$

$$CO(g/mi)= \left[\frac{0.43*CO(C)+1.0*CO(S)+0.57*CO(H)}{7.5}\right] \quad (2)$$

$$NO_x(g/mi)= \left[\frac{0.43*NO_x(C)+1.0*NO_x(S)+0.57*NO_x(H)}{7.5}\right] \quad (3)$$

Likewise, the urban fuel economy is computed as follows:

$$MPG_u= \left[\frac{7.5}{0.43*F(C)+1.0*F(S)+0.57*F(H)}\right] \quad (4)$$

The highway fuel economy is simply the miles driven during the highway portion divided by the gallons consumed:

$$MPG_h = \left[\frac{10.2}{F(Hwy)}\right] \quad (5)$$

Equations (4) and (5) are combined (as prescribed by the EPA) to form the composit fuel economy:

$$MPG_{composite} = 100/\ (55/MPG_u) + (45/MPG_h) \quad . \quad (6)$$

The denominator of Equation (6) represents the weighted sum of the fuel consumed over the four distinct segments of the test. The weighting function is called $B(t)$. Integrating the product of $B(t)F(t)$ over the interval $[0,t_f]$, results in the denominator of Equation (6). Where $\dot{F}(t)$ is the fuel rate and t_f is the length of the test in seconds. Maximization of the composite MPG is equivalent to minimizing the integral of $B(t)\dot{F}(t)$. In the next section, the optimal automotive engine control problem is stated as a minimum fuel problem using this integral. Similarly, if A(t) is the weighting function of the emissions, then the integrals of the products of A(t) and the individual emission rates are the cumulative emissions over the entire test (in gram/mile).

Necessary Conditions of Optimality and the Emission Constraints

The optimal automotive engine control problem, which is qualitatively described in the preceding section, can be precisely formulated as follows.

$$\text{MINIMIZE:} \quad J = \int_0^{t_f} L(\underline{u},t)dt \tag{7}$$

subject to the terminal constraints:

$$\begin{aligned} x_1(t_f) &\leq HC^* \\ x_2(t_f) &\leq CO^* \\ x_3(t_f) &\leq NO_x^* \end{aligned} \tag{8}$$

and to the control constraint:

$$C(\underline{u},t) \leq C^*(t) \text{ for all t on the interval } [0,t_f]. \tag{9}$$

In the above equations, L is the weighted fuel consumption rate. The three components of the $\underline{x}$ vector are the accumulated mass emissions of hydrocarbons, carbon monoxide, and oxides of nitrogen, respectively. The fuel consumption rate and emission rates are weighted as described earlier. HC*, CO*, and NO_x* are the specified mass emission limits. The time at the end of the test, t_f, is fixed. C*(t) is the specified driveability limit, and $C(\underline{u},t)$ is the driveability at time t. The control vector $(A/F,SA,EGR)^T$ is denoted by $\underline{u}(t)$, where $(\ )^T$ designates the transpose of the matrix ( ).

The fuel rate, L, the emission rates, $\dot{x}_i$, and the driveability, C, are functions of the controls (A/F,SA,EGR) as well as engine load, speed and temperatures. However, if the driving schedule is specified, as with the FTP, then for a given system the engine load, speed, etc., are implied as functions of time from the start of the test. Therefore, the fuel rate, the emission rates, and the driveability can be expressed as functions of the controls and of time from the start of the test:

$$\begin{aligned} L &= L\ (A/F,\ SA,\ EGR,\ t) & & \\ \dot{x}_1 &= f_1\ (A/F,\ SA,\ EGR,\ t) & x_1(0) &= 0 \\ \dot{x}_2 &= f_2\ (A/F,\ SA,\ EGR,\ t) & x_2(0) &= 0 \\ \dot{x}_3 &= f_3\ (A/F,\ SA,\ EGR,\ t) & x_3(0) &= 0 \\ C &= C\ (A/F,\ SA,\ EGR,\ t) & & \end{aligned} \tag{10}$$

The implications of this formulation are of fundamental importance in this approach to the optimal automotive engine control problem and have particular significance in its practical application.

First, consider the emission constraints. If the terminal inequality constraints are treated as equality constraints, a standard Bolza problem results and the Minimum Principle can be employed. The Hamiltonian function, H, is defined:

$$H = L + p_1f_1 + p_2f_2 + p_3f_3, \tag{11}$$

where the Lagrange multipliers satisfy the Euler-Lagrange equations:

$$\begin{aligned} \dot{p}_1 &= -\frac{\partial H}{\partial x_1} = 0 \\ \dot{p}_2 &= -\frac{\partial H}{\partial x_2} = 0 \\ \dot{p}_3 &= -\frac{\partial H}{\partial x_3} = 0 \end{aligned} \tag{12}$$

where ($\dot{\ }$) denotes the derivative of ( ) with respect to time. Since neither L nor $\underline{f}$ are functions of the state ($\underline{x}$) the partial derivatives of the Hamiltonian (H) with respect to the state variables are, therefore, zero. Hence, by the Euler-Lagrange equations, the multipliers are constant functions. The Hamiltonian function is discontinuous at the discontinuities of the system equations:

$$\begin{aligned} L(\underline{u},t) &= B(t)\dot{F}(\underline{u},t) \\ f_1(\underline{u},t) &= A(t)\dot{HC}(\underline{u},t) \\ f_2(\underline{u},t) &= A(t)\dot{CO}(\underline{u},t) \\ f_3(\underline{u},t) &= A(t)\dot{NO}_x(\underline{u},t) \end{aligned} \tag{13}$$

because of the jumps in the weighting functions A(t) and B(t). However, this does not affect the computational solution because the final time is fixed.

This problem structure has certain computational advantages. The fact that the multipliers are continuous and constant not only simplifies the involved calculations, but it also makes practical the direct calculation of the multiplier values necessary to satisfy the terminal constraints on HC, CO, and NO_x when gradient-type methods are employed. For example, let $HC(t_f)$, $CO(t_f)$, and $NO_x(t_f)$ be the emissions at the end of the test resulting from an arbitrary control vector representing the reduction in the three emissions constituents necessary to meet the constraints, i.e.,

$$\underline{\Delta x_f} = \begin{bmatrix} HC^*-HC(t_f) \\ CO^*-CO(t_f) \\ NO_x^*-NO_x(t_f) \end{bmatrix} . \tag{14}$$

Let $\underline{\delta u}(t)$ denote a variation of the control vector,

$$\underline{\delta u}(t) = \begin{bmatrix} A/F(t) - A/F\big|_{ref}(t) \\ SA(t) - SA\big|_{ref}(t) \\ EGR(t) - EGR\big|_{ref}(t) \end{bmatrix} \quad (15)$$

To satisfy the constraints it is necessary that the required change in the emission constituents, $\underline{\Delta x_f}$, be equated to that resulting from the variation in the control, i.e.,

$$\underline{\Delta x_f} = \int_0^{t_f} \left\{ \left[\frac{\partial f}{\partial u}\right]_{ref} \underline{\delta u} + \left[\frac{\partial f}{\partial x}\right]_{ref} \underline{\delta x} \right\} dt, \quad (16)$$

where

$$\underline{\dot{\delta x}} = \left[\frac{\partial f}{\partial u}\right]_{ref} \underline{\delta u} + \left[\frac{\partial f}{\partial x}\right]_{ref} \underline{\delta x}, \underline{\delta x}(0)=0. \quad (17)$$

However, since the emission rates are not functions of the states (the accumulated emission), i.e., $\underline{x}$ does not appear in $\underline{f}(\underline{u},t)$, the matrix, $\left[\frac{\partial f}{\partial x}\right]$, is identically zero. If a gradient-type method is employed, then

$$\underline{\delta u} = -\left[K\right]\left[\frac{\partial H}{\partial u}\right]^T_{ref} = -\left[K\right]\left[\frac{\partial L}{\partial u}\bigg|_{ref} + \underline{p}^T \frac{\partial f}{\partial u}\bigg|_{ref}\right]^T \quad (18)$$

Therefore,

$$\underline{\Delta x_f} = -\int_0^{t_f} \left\{ \left[\frac{\partial f}{\partial u}\right]\left[K\right]\left[\frac{\partial L}{\partial u}\right]^T + \left[\frac{\partial f}{\partial u}\right]\left[K\right]\left[\frac{\partial f}{\partial u}\right]^T \underline{p} \right\}_{ref} dt \quad (19)$$

If $\underline{f}(\underline{u},t)$ and $L(\underline{u},t)$ are known then it is possible to solve this equation for the vector of Lagrange multipliers:

$$\underline{p} = -\left[\int_0^{t_f} \left\{ \left[\frac{\partial f}{\partial x}\right]\left[K\right]\left[\frac{\partial f}{\partial u}\right]^T \right\}_{ref} dt\right]^{-1} \left[\underline{\Delta x_f} + \int_0^{t_f} \left\{ \left[\frac{\partial f}{\partial u}\right]\left[K\right]\left[\frac{\partial L}{\partial u}\right]^T \right\}_{ref} dt\right] \quad (20)$$

Thus, the Lagrange multipliers can be calculated directly from the above equation, and the calculation requires only one forward integration of the system. (The usual backward integration for the Lagrange multipliers is avoided because of the problem formulation.)

Furthermore, since it is possible to calculate the Lagrange multipliers associated with any nominal trajectory, a gradient procedure can be defined to treat the inequality constraints directly. In the Lagrange form of the problem with specified terminal states, $\underline{p}(t_f)$ is the gradient of the cost J with respect to the specified states. Writing Equation (8) as follows:

$$h_i\left[\underline{x}(t_f)\right] = \left[x_i(t_f) - x_i^*\right] \le 0 \quad (21)$$
$$i = 1,2,3$$

and since Equation (8) represents an inequality constraint on each of the terminal states, it is necessary (L1) that:

$$p_i(t_f) \ge 0$$
$$p_i(t_f) h_i\left[\underline{x}(t_f)\right] = 0 \quad (22)$$
$$i = 1,2,3$$

Consequently, if Equations (20) and (22) are inconsistent, i.e., Equation (20) specifies a $p_i<0$, then the terminal inequality constraint on the ith state is inactive and that p_i is set equal to zero. The remaining multipliers are computed by deleting the ith component of the $\underline{\Delta x_f}$ vector and the ith row from the $\left[\frac{\partial f}{\partial u}\right]$ matrix in Equation (20). The improved estimate of the control vector is specified by adding Equation (18) to the reference control.

Development of a Driveability Constraint

A measurement of driveability during transient driving is developed by this author in (D6). This driveability measurement is employed in the experiments described later. During the experiments two different schemes were employed to enforce the driveability constraint, Equation (9), in the optimization computational algorithm. The rate of convergence and iteration stability of these two schemes are compared in the next section. The two schemes are:

1) the influence function approach, and
2) the interior-exterior penalty function approach.

The Influence Function Approach. In this approach the control adjustment, Equation (18), is modified whenever the driveability constraint, Equation (9) is violated by the reference trajectory. The resulting equation is:

$$\underline{\delta u}(t) = -\left[K\right]\left[H_u + \mu(t) C_u\right]^T_{ref} \quad (23)$$

Where the subscript u denotes the gradient of that function with respect to the control. The value of the influence function, $\mu(t)$, is determined to enforce the constraint. That is, by equating the first order terms of the Taylor Series expansion in the controls to the excess of Equation (9) and substituting Equation (23) for $\underline{\delta u}$:

$$(C - C^*) = -\left\{\left[C_u\right]\left[K\right]\left[H_u + \mu(t) C_u\right]^T\right\}_{ref} \quad (24)$$

Equation (24) can be solved for $\mu(t)$:

$$\mu(t) = \frac{1}{C_u K C_u^T}\left[(C-C^*) - C_u K H_u^T\right] \tag{25}$$

Let $g(\underline{u},t) = (C-C^*)$, then the influence function, $u(t)$, must satisfy the following conditions:

$$\mu(t)\; g(\underline{u},t) = 0 \tag{26}$$

$$\mu(t) \geq 0$$

An inconsistency between Equations (25) and (26) implies the constraint, Equation (9) is inactive. For the case that Equation (25) specifies $\mu<0$, Equation (18) would tend to force the controls back into the admissible region. The scaler $\mu(t)$ is set equal to zero wherever the reference trajectory satisfies the constraint. Because the constraint is not a function of the state, the Lagrange multipliers associated with the emissions remain constant on the constraint boundary $\left\{C(\underline{u},t) - C^*(t)\right\} = 0$.

<u>The Interior-Exterior Penalty Function Approach.</u> This approach modifies the performance criterion (Equation (7) as follows:

$$\text{MINIMIZE:}\quad J = \int_0^{t_f}\left[L(\underline{u},t) - W^{-1}C^*(t)\,(C(\underline{u},t) - C_p(t))^{-1}dt\right] \tag{27}$$

where W is a weighting factor determined experimentally to balance the control adjustments due to fuel economy and emissions against those due to the driveability considerations, such that neither dominates to the exclusion of the other. The $C_p(t)$ is set equal to 2 $C^*(t)$. This form of penalty function, $W^{-1}C^*(t)\;(C(t) - C_p(t))^{-1}$, is selected to provide a continuously varying influence of the driveability constraint on the control adjustments; less when the constraint is satisfied, more when it is violated. This characteristic compares with the binary situation that exists with the influence function approach in that it tends to make the interior-exterior penalty function approach more stable than the influence function approach.

In Figure (1) the behavior of the penalty function for various driveability measurements is shown. Note that the influence exerted interior to the constraint (i.e., $C<C^*$) is minor compared to the influence exerted exterior to the constraint (i.e., $C>C^*$). To avoid computation problems near $C=C_p$, $C^*(t)\;C(t)-C_p(t)^{\;-1}$ is limited to a value of four for any $C(t)>(C_p(t)-0.25C^*(t))$ as shown by the dashed line in Figure (1).

With the performance criterion, Equation (27), the Hamiltonian is:

$$H = L - W^{-1}C^*(t)(C-C_p)^{-1} + p_1 f_1 + p_2 f_2 + p_3 f_3 \tag{28}$$

The corresponding expressions for the Lagrange multiplier vector and the control adjustment are, respectively:

$$\underline{p} = -\left[\int_0^T\left\{\left[\frac{\partial f}{\partial u}\right]\left[K\right]\left[\frac{\partial f}{\partial u}\right]^T\right\}_{ref} dt\right]^{-1} \cdot \left[\Delta\underline{x}_f + \int_0^T\left\{\left[\frac{\partial f}{\partial u}\right]\left[K\right]\left[\left[\frac{\partial L}{\partial u}\right] + W^{-1}C^*(t)(C-C_p)^{-2}\left[\frac{\partial C}{\partial u}\right]\right]^T\right\}_{ref} dt\right] \tag{29}$$

$$\delta u(t) = -\left[K\right]\left[\frac{\partial H}{\partial u}\right]^T_{ref} = -\left[K\right]\left[\frac{\partial L}{\partial u} + W^{-1}C^*(t)(C-C_p)^{-2}\frac{\partial C}{\partial u} + \underline{p}^T\frac{\partial f}{\partial u}\right]^T_{ref} \tag{30}$$

Equations (23) - (30) define two different approaches to the solution of the optimal automotive engine control problem with the driveability constraint. The influence function approach is more direct, but it also tends to be less stable when resolving the tradeoffs between fuel economy, emission, and driveability which tend to be diametrically opposed. The interior-exterior penalty function approach, though somewhat indirect, tends to be more stable. The influence function can be desensitized by employing under-relaxation techniques at the expense of the rate of convergence. However, as is shown later, the interior-exterior penalty function approach appears to provide the best trade-off between rate of convergence and stability.

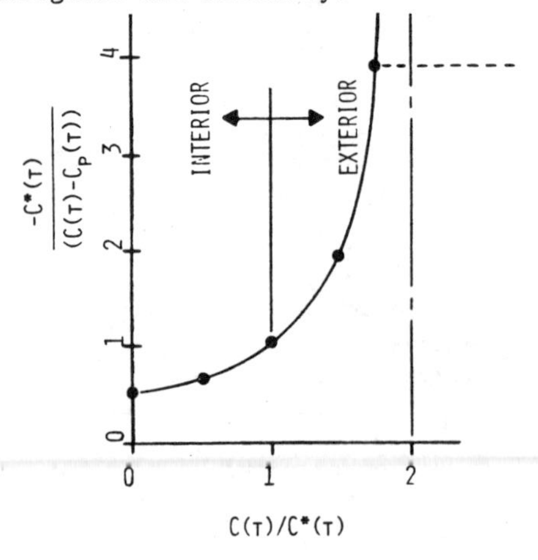

Fig. 1 Interior-Exterior Penalty Function

Definition of a Gradient-Type Algorithm

Realistic mathematical models for fuel economy, emissions, and driveability have not been developed for transient system performance. Therefore, to take advantage of the structure of the optimal control problems described above while maintaining an accurate representation of the physical processes, one must develop all of the required information by experimental means. This, of course, implies that the entire system (engine, vehicle, catalytic converter, and controller) must be implemented and driven over the test driving schedule to experimentally evaluate the system equations. Futhermore, these experimental evaluations are relatively expensive in terms of facilities and test time; and , therefore, any performance criterion of an iterative procedure employed to solve this problem is most likely based on its efficiency in terms of number of evaluations of the system equations.

In the remainder of this section, a gradient-type procedure is defined which utilizes the advantageous structure of the problem developed in the preceding sections. Because the state equations are only functions of the controls and time, Equation (20) or (29) is explicitly evaluated for the Lagrange multipliers thus eliminating the q integrations of the system equations needed to determine the transistion matrix, $\frac{\partial x(t_f)}{\partial p(t_0)}$ (B2) (where q is the number of active constraints). Furthermore, as described earlier the procedure directly addresses terminal inequality emission constraints.

A First-Order Gradient-Type Algorithm

Step 1. Estimate control variable history, $\underline{u}(t)$.

Step 2. Experimentally integrate the system equations with respect to $\underline{x}(0) = 0$ and the control variable histories from Step 1, during the integration record L(t), C(t), and $\underline{x}(t_f)$.

Step 3. Repeat Step 2 m-times (m is the dimension of the control vector $\underline{u}(t)$), each time perturbing one of the components of $\underline{u}(t)$.

Step 4. Numerically compute $\left[\frac{\partial L}{\partial u}\right]$, $\left[\frac{\partial f}{\partial u}\right]$, and $\left[\frac{\partial C}{\partial u}\right]$ from the results of Steps 2 and 3.

Step 5. Integrate the coefficients of the Lagrange multipliers (Equation (22) or (27) using the results of Step 4).

Step 6. Evaluate the Lagrange multipliers and check for inactive constraints ($p_i<0$). If one (or more) of the emission constraints is (are) inactive, then set the corresponding multipliers equal to zero and recompute the remaining multipliers.

Step 7. Evaluate the improved estimate of of the control $u^{(i+i)} = u^{(i)} + \delta u^{(i)}$, where $\delta u^{(i)}$ is specified by Equation (18), (23) or (30), and repeat Steps 1 through 7 until $|\Delta \underline{x}(t_f)| \leq \varepsilon_1$ and $| H_u^{i+i} - H_u^{i} | \leq \varepsilon_2$, where $\varepsilon_1, \varepsilon_2 > 0$ are prespecified.

In this section, it has been assumed that the underlying assumptions of the Minimum Principle hold and that the system equations can be evaluated experimentally. In (D6), the operational aspects are considered in more detail and the underlying assumptions are validated. Difficulties associated with transient vehicle modelling are also discussed in (D6). These difficulties provide the motivation for the automotive certification engineer to generate the various year-to-year engine calibrations experimentally. Similarly, to optimize an engine calibration, one is motivated to attempt the experimental approach. In the following section, this procedure is applied to an experimental engine/vehicle system to develop an improved calibration.

EMISSION-CONSTRAINED MINIMUM FUEL PROBLEM

In the previous section, a gradient-type algorithm is defined for solving the emission-constrained minimum fuel problem. In reference (D6), the experimental setup is described. In this section the theoretical results described earlier are combined with the experimental hardware of (D6) to solve the emission-constrained minimum fuel problem over the 1978 FTP driving schedule.

Since this is the first time that optimal control theory is applied to the engine control problem employing the actual system hardware, as opposed to a complete mathematical model, the problem is solved in sequentially more difficult stages to develop insight into the approach. Each of these stages illustrates a particular aspect of the optimization procedure. The resultant problems and their experimental solution are described in detail in the remainder of this paper.

Operational Considerations

This section addresses three issues concerning the experimental application of the gradient procedure defined earlier.

Feedback Control Function. The gradient procedure manipulates the time history of the control vector to converge to the optimal solution of the particular problem, i.e., the optimal time history of the control. Two significant difficulties arise when attempting to deal with the time histories of the control vector with the experimental setup. First, storing the three components of the control vector for every half-second interval over the complete FTP requires 15,852 data points. This exceeds the storage capacity of the engine control minicomputer without consideration of the program storage requirements. Furthermore, it is important to maintain the feedback relationship between the engine operating conditions and the commanded controls. Small phasing differences between the controls, particularly spark advance, and the engine operating variables (speed, load, etc.) result in significant differences in end-of-test fuel economy and emissions. These difficulties are circumvented by parameterizing the time history of the controls in terms of feedback control functions. That is, by initially estimating the feedback control functions and programming the engine control minicomputer accordingly, the time history is generated as the control functions are evaluated during the FTP. The perturbation runs are still used to compute the gradients, which in turn are used to compute the control adjustments for each half-second interval during the test. The time history of the control adjustments are correlated with the time history of the feedback variables to form feedback control adjustment functions as shown in Figure 2. These feedback adjustment functions are added to the previous feedback control functions and the procedure is repeated. The modified gradient procedure is shown in Figure 3.

<u>The Weighting Matrix (K).</u> In determining the weighting matrix, (K), for the gradient step, one must provide two types of information. First, the relative weighting among the components of the gradient vector; and second, the overall stepsize. For this purpose, let:

$$(K) = \alpha\,(K^*) \tag{31}$$

The matrix (K*) is employed mainly to provide proper scaling for the problem. By assuming the multipliers remain unchanged and writing the second-order expansion of J about the reference run:

$$\delta J = \int_0^{t_f} \left[H_u\, \delta_u + \tfrac{1}{2}\, \delta u^T H_{uu}\, \delta u \right] dt_{ref} \tag{32}$$

an appropriate (K*) can be specified. The control variation is determined by the condition:

$$\frac{dJ}{d(\delta u)} = H_u + \delta u^T H_{uu} = 0 \tag{33}$$

from which it is clear that the minimum occurs if:

$$\delta u = -H_{uu}^{-1}\, H_u^T \tag{34}$$

Thus, by assuming (K*) to be diagonal, and by specifying:

$$K^*_{11} = 1$$
$$K^*_{22} = \left[\frac{\partial^2 H}{\partial (A/F)^2}\right]\left[\frac{\partial^2 H}{\partial (SA)^2}\right]^{-1} \tag{35}$$
$$K^*_{33} = \left[\frac{\partial^2 H}{\partial (A/F)^2}\right]\left[\frac{\partial^2 H}{\partial (EGR)^2}\right]^{-1}$$

a crude Newton-type algorithm results. The second partial derivatives of the Hamiltonian were experimentally evaluated and the average valves of the ratios specified:

$$K^*_{11} = 1$$
$$K^*_{22} = 50 \tag{36}$$
$$K^*_{33} = 1$$

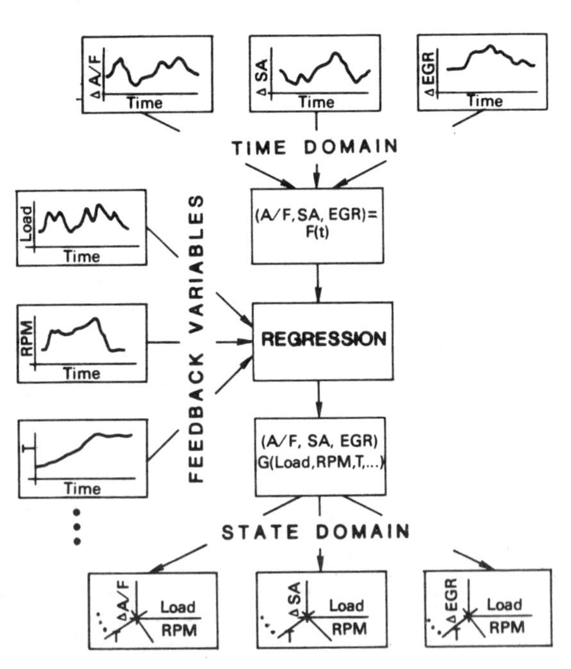

Fig. 2. Regression of Controls with Feedback Variables.

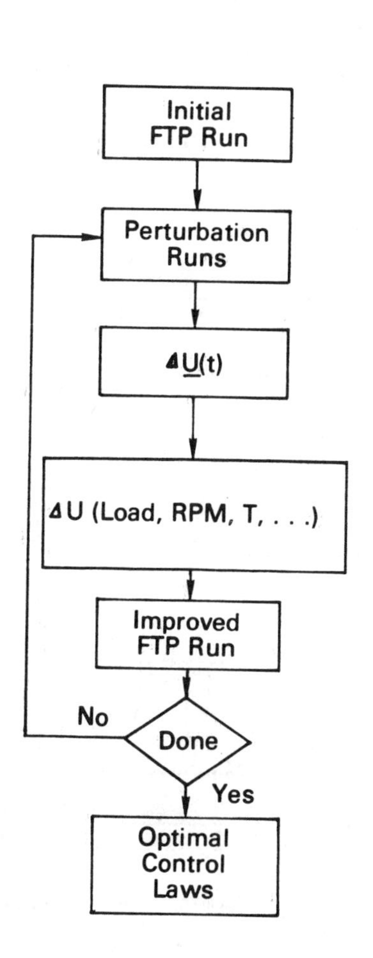

Fig. 3. Experimental Gradient Procedure.

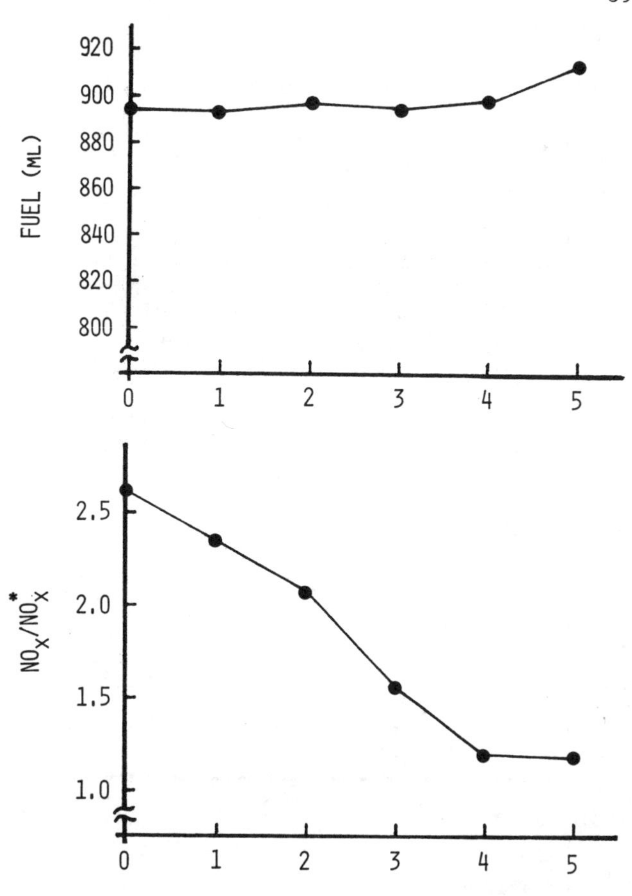

Fig. 4. Fuel & Emission Versus Stepsize

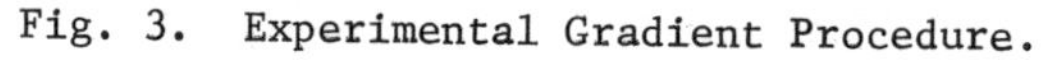

The stepsize, α , is determined during the first iteration by performing a crude linear search. The fuel consumption and emission results versus stepsize are shown in Fig. 4. for the hot-start optimization problem of the next section. The NO_x decreases monotonically with only a minor increase in fuel consumption for values of $\alpha \leq 4$. For α=5, the increase in fuel consumption is noticeably larger and the decrease in NO_x considerably smaller. Therefore, α=4 appears to be a reasonable stepsize. Hence, the weighting matrix is specified:

$$(K) = \alpha(K^*) = \begin{bmatrix} 4 & 0 & 0 \\ 0 & 200 & 0 \\ 0 & 0 & 4 \end{bmatrix} \qquad (37)$$

Other checks are also included to avoid the influence of any suspect data. Any indicated adjustments larger in magnitude than three times the sample standards deviation of the adjustment over the entire test are ignored. Checks are also included to monitor how many data points are ignored and for what reason.

Minimum Fuel Problem With Emission Constraints

The first problem to be considered employing this optimization procedure is that of minimizing fuel consumption over the hot five-cycle portion of the federal fuel economy and emissions test. Air-fuel ratio and spark advance are the controls; no external EGR is considered. Since only the hot-start portion of the test is considered, the hydrocarbons and carbon monoxide are not of primary concern. Since only a portion of the FTP is being considered, the emission constraints must be chosen based on previous experience. The oxides of nitrogen constraint limit is chosen to correspond to a 2.0 g/mi standard over the entire 1978 FTP.

The formal driveability constraint, defined in (D6) is not considered here, but is imposed on this problem later. Note that although no formal driveability constraint is imposed during the optimization described in this section the procedure has an implicit driveability-type requirement in that the engine and automatic driver combination must behave in a repeatable manner throughout every test. This feature could be considered as a crude, first-order driveability constraint which is intrinsically imposed by the procedure described in this thesis. Typically this is not the case with mathematical model or steady-state oriented optimization schemes, as has been revealed when their results are implemented.

To begin the optimization procedure, initial estimates of the feedback control functions must be specified. The initial air-fuel ratio and spark advance functions are shown in Fig. 5. The initial estimate of the air-fuel ratio function is a constant 17:1 for all engine loads and speeds. This air-fuel ratio corresponds closely to that for minimum brake specific fuel consumption (BSFC) for the engine. The initial spark advance function is the production distributor curve mapped from manifold vacuum to the fueling function at an air-fuel ratio of 17:1. The typical speed-load trends are evident: the spark timing advances as the engine speed increases, and the spark timing retards as the engine load increases.

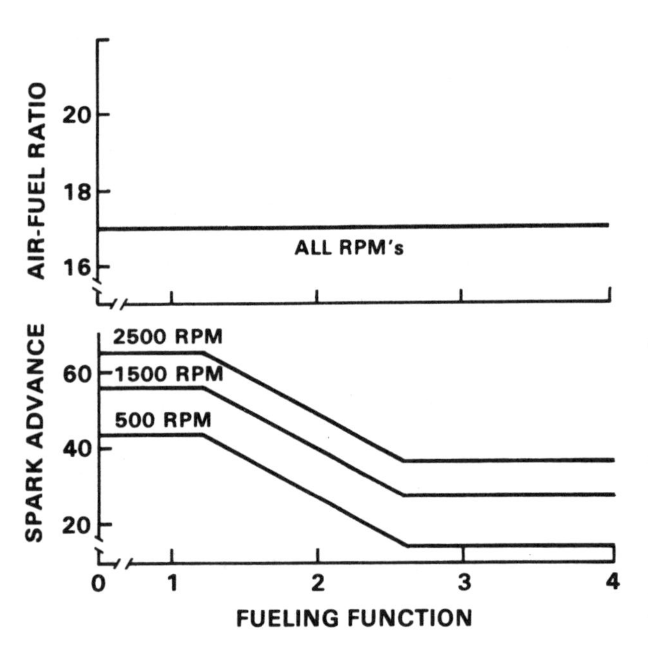

Fig. 5. Initial Estimates of Hot-Start Feedback Control Functions.

The reference and the two perturbation controls are run to compute the gradients of the fuel and emission rates with respect to the control vector. Thus, the gradient of the Hamiltonian can be evaluated and weighted to specify a control variation which improves the fuel and/or emissions. The control variation for the second cycle of the hot-start test is shown in Figure 6. The vehicle speed trace is shown in the uppermost figure, while the adjustments to the air-fuel ratio and the spark advance are shown below. Note that these adjustments are incorporated into the control functions for subsequent iterations. Therefore, as the procedure converges, the indicated adjustments uniformly vanish for all time during the test. If the indicated adjustments do not vanish, additional control function modifications must be made.

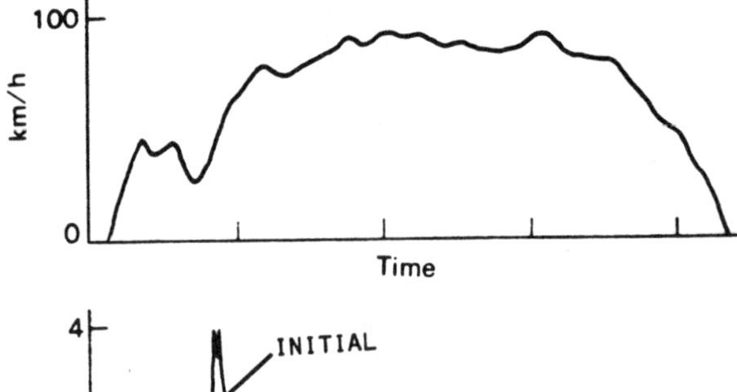

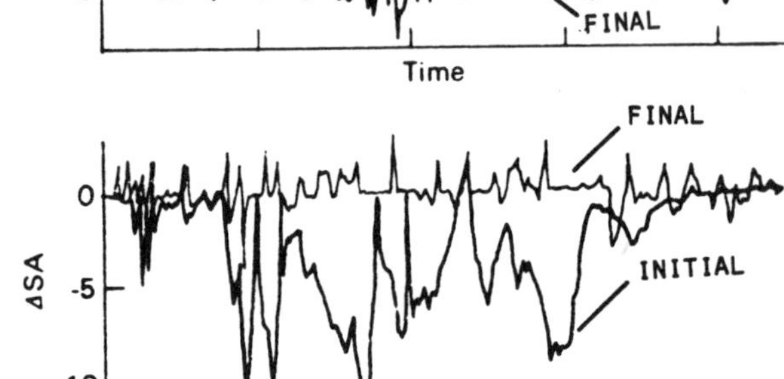

Fig. 6. Control Adjustments Versus Time For the Second Cycle

The first-iteration adjustments, which are calculated as functions of time, must subsequently be quantified as functions of the feedback variables input to the controllers that operate the engine for a particular driving schedule. For this problem, the feedback variables are the fueling function and the engine speed. A multivariant regression analysis provides a somplete second-order expansion of the control adjustments in the feedback variables. The adjustment functions are programmed into the engine control minicomputer and combined with the feedback control functions from the current iteration to form the control functions for the reference run of the next iteration.

After two iterations, the procedure converges to the emission constraint to provide the feedback control functions which are shown in Figure 7. The influence of both load and speed appear in the optimal feedback functions. A third iteration provides no further improvement in fuel economy, so the problem is assumed to be converged.

It is of interest to compare the control adjustments in Figure 6. The second iterate adjustments for both air-fuel ratio and spark advance are reduced significantly over the engine test compared to the adjustments from the first iteration. Had it not been possible to reduce these adjustments uniformly

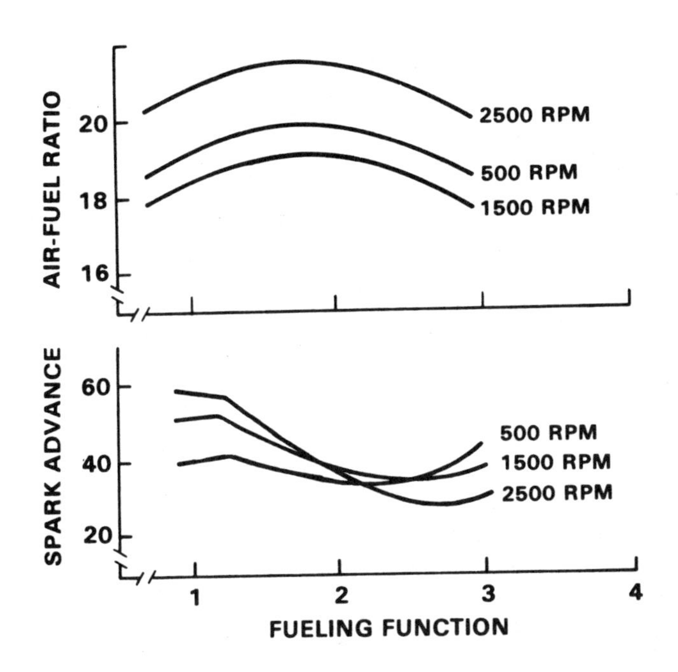

Fig. 7. Optimal Hot-Start Feedback Control Functions

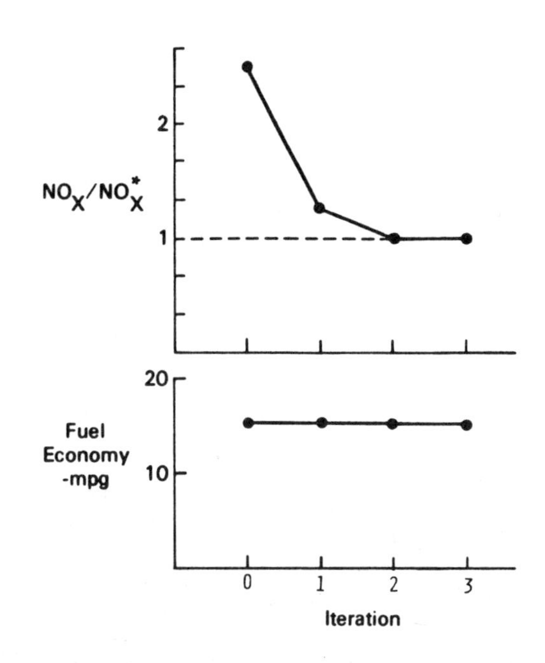

Fig. 8. Hot-Start Emissions & Fuel Economy

Throughout the test then further improvements could only be made by including more parameters in the feedback functions. Since the indicated adjustments are reduced to such small levels, the current parameterization is adequate for this problem.

The fuel economy and the NO_x emissions after each iteration of the procedure are shown in Figure 8. The initial choice of the control functions did not satisfy the NO_x constraint; in fact, the NO_x from the initial estimates of the control function is over 2.5 times the constraint. However, in two iterations, the NO_x is decreased to the constrained value with a very minor penalty in fuel economy. Since this problem involves only the hot-start portion of the test with an oxidizing converter, the carbon monoxide and hydrocarbons start and remain below the emission constraint values for all iterations.

It is interesting to compare optimal feedback control functions (Figure 7) to the initial estimates (Figure 5) and interpret the effected changes. The reduction in the NO_x emissions is achieved primarily by the overall leaning of the air-fuel ratio, particularly during the high-speed, medium-load cruises encountered in the second cycle. At the heavier load conditions, additional spark advance is provided to balance the effective spark retard caused by leaning the air-fuel ratio.

The Cold-Start Engine-Catalyst Warmup Problem

The cold-start optimization problem, catalytic converter warmup, and transient choke problems have not been thoroughly treated in the literature. The most sophisticated treatment of transient catalytic converter effects on emissions has been the development of a matrix of pass-through ratios for various loads and speeds for fully warm operation. In actuality the efficiency of the converter is dependent upon its temperature and the control values as well as engine speed and load. The emissions produced during the cold-start portion of the test, as the engine and catalyst warmup, represent up to 80% of the total HC and CO for the complete test. Thus, to treat the HC and CO constraints realistically, an optimization scheme must consider the cold-start portion of the test, including the engine enrichment function (choke) and the cold, warm, and hot dynamics of the catalytic converter.

To demonstrate the ability of this optimization procedure to handle the engine-converter warmup, fuel consumption is minimized during the five-cycle cold-start portion of the test. Air-fuel ratio and spark advance are the available controls and the emissions are measured downstream of the catalytic converter. The target emission constraints are chosen to correspond to:

$HC\ (t_f) \leq 0.3$ g/mi

$CO\ (t_f) \leq 9.0$ g/mi

$NO_x(t_f) \leq 2.0$ g/mi

for the 1978 FTP.

The initial estimates of the feedback control functions for air-fuel ratio and spark advance are shown in Figure 9a-b. As in the preceding section, the initial estimate of the air-fuel ratio function is a constant 17:1. However, for the cold-start test, it is necessary to add a power enrichment-type function to enable the cold engine to make the acceleration to 55 MPH during the second cycle of the test. There is no dependence on engine speed or temperature in the initial estimate of the air-fuel ratio control function. The initial estimate of the spark advance control function is similar to that in the preceding section shown in Figure 5.

Operation of the engine employing these control functions over the FTP maps the control functions into the cumulative end-of-test fuel consumption and emissions. As expected, the initial estimates of the control functions do not satisfy the emission constraints. As in the previous problem, after the emission test is run with the initial estimates of the control functions, each of the controls is perturbed individually. (The air-fuel ratio is shifted one ratio throughout the test for the first perturbation run; likewise the spark advance is shifted five degrees throughout the second perturbation run.) Then the gradients of the fuel rate and emission rates are numerically computed from the half-second scanning data. With these gradients and the excess emissions, Δx_f, Equation (20) is evaluated to obtain the multiplier values. Once the multiplier values are known, Equation (18) prescribes the control adjustments for every half second throughout the test. These adjustments, which are functions of time, are correlated with the feedback variables to form the adjustment feedback functions. The adjustment feedback functions are added to the feedback functions of the reference fun for the following iterations.

In this experiment the HC and NO_x constraints, along with the fuel consumption, are the active influences on the optimal control feedback functions. The CO constraint is inactive; thus, $p_2 = 0$ by Equation (22). The remaining multipliers are computed as described previously.

The vehicle speed and the calculated adjustments to the controls are shown for the second cycle of the test in Figure 10. The control adjustments are prescribed by Equation (18). The solid lines show the adjustments from the first iteration, and the dashed lines show the adjustments from the final (i.e., second) iteration. As expected, the magnitudes of the indicated adjustments are reduced uniformly throughout the test from the first iteration to the final iteration.

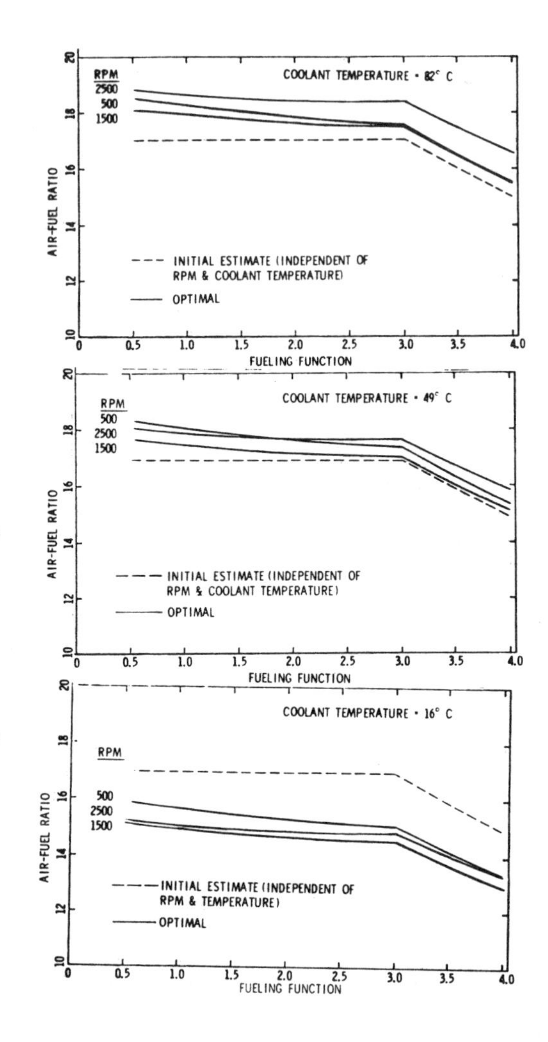

Fig. 9a. Cold-Start Feedback Functions For Air-Fuel Ratio

Peaks that occur in the adjustment functions that are not reduced by subsequent iterations indicate the inability of the feedback variables to represent the optimal control functions during a particular mode of operation. Thus, a structural change in the controller is required. The majority of these peaks occur before the engine has reached its designed operating temperature, hence, these peaks are assumed to be associated with the warmup transient. Since a coolant temperature signal is available in the experimental hardware, it is employed as a feedback variable to characterize the system

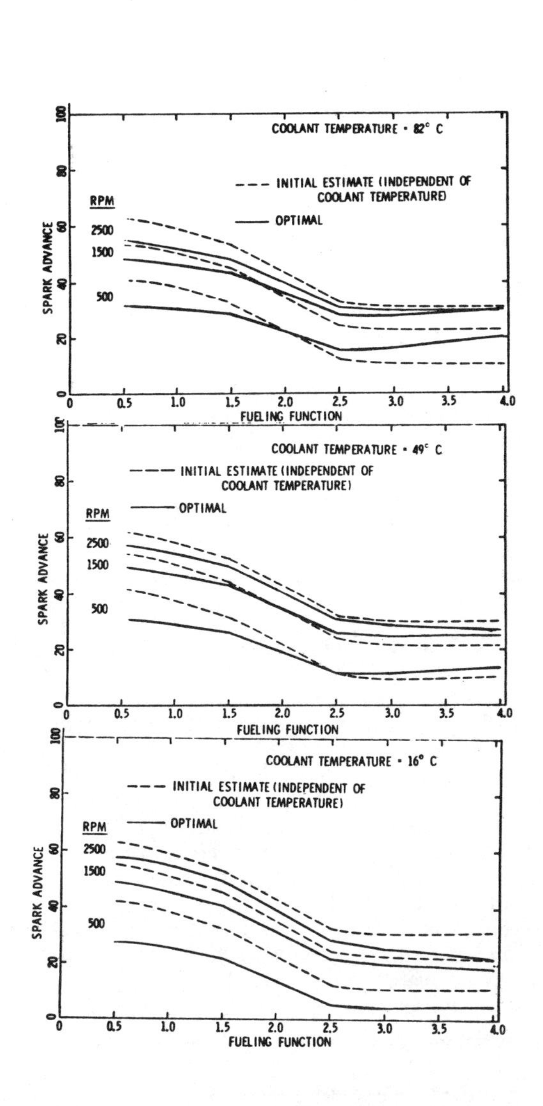

Fig. 9b. Cold-Start Feedback Functions For Spark Advance

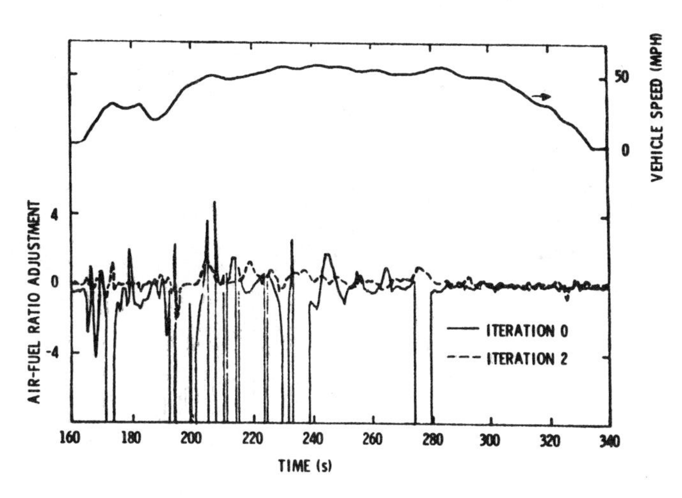

Fig. 10a. Air-Fuel Ratio Adjustments Versus Time For The Second Cycle

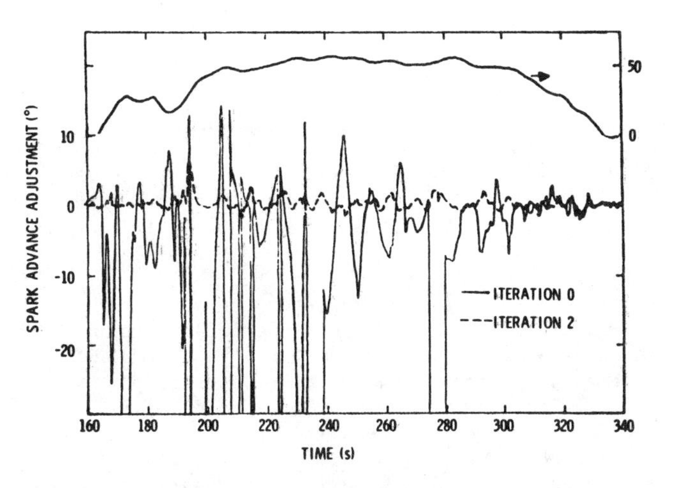

Fig. 10b. Spark Advance Adjustments Versus Time For The Second Cycle

operating temperature during the warmup transient. With the inclusion of coolant temperature as a feedback variable, these peaks decrease on the following iteration.

The initial estimates of the control feedback functions are augmented to include the coolant temperature effects, as shown in Figures 9a-b. Likewise, the spark advance control function is augmented during the optimization as shown. The resultant normalized emissions and fuel economy after each iteration are shown in Figure 11. The normalized emission constraints are shown as unity. The fuel economy is improved by more than 8% while the HC is decreased by more than 50% and the NO_x is decreased by more than 35%.

A few brief comments about these results are in order. First, Figure 9a. shows the optimal cold-start air-fuel ratio is approximately 15:1 at 60^{0}F coolant temperature. This is, of course, before catalyst light-off and before the exhaust system is hot enough to permit any HC oxidation in the exhaust manifold. During this initial warmup period the primary driving force on the air-fuel ratio appears to be the maintenance of low HC levels until the converter lights off. The optimal air-fuel ratio changes as the exhaust system warms up and more oxidation occurs in the exhaust manifold. The trade-off appears to be between minimizing the engine-out HC after initial start-up and providing leaner mixtures to take advantage of the oxidation in the exhaust manifold and catalyst as they warm up.

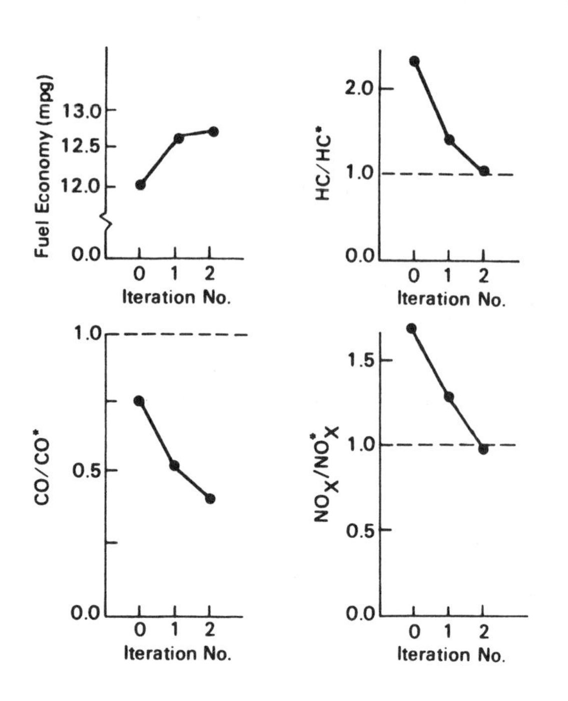

Fig. 11. Cold-Start Emissions & Fuel Economy

The optimal spark advance for the experimental system is more retarded at the cooler engine temperatures. This is contrary to some contemporary production systems, which provide additional spark advance for a cold engine. The difference is probably due to the structure of the controller, as discussed in (D6). The experimental system is an air-control system in which the air throttle response to a change in fuel rate is damped at cooler engine temperatures. This is done in such a way that the changes in air flow into the cylinder are phased with the changes in fuel flow into the cylinder. At cooler engine temperatures, the fuel volatility, wall wetting, and fuel transport through the intake manifold significantly affect changes in the fuel flow at the cylinder compared to the behavior at the inlet to the intake manifold.

In a carburetted system, or in any "fuel control" system, as opposed to "air control", the air-fuel mixture tends to go lean with a sudden increase in engine throttle opening, particularly with a cold engine. Fuel control systems which employ additional cold-start spark advance provide spark timing which is a compromise between that which is appropriate at the calibrated quasi-steady-state air-fuel ratio and that which is appropriate during the lean excursions that occur during acceleration. However, with this experimental air-control system it is possible to operate with more spark retard, for HC control at cooler engine temperatures, than with the conventional fuel control system. This is because of the appropriate phasing of air flow with fuel flow at the engine cylinder.

Minimum Fuel Problem With Emission and Driveability Constraints

In this section the minimum fuel problem discussed earlier is generalized to include an explicit driveability constraint. The driveability constraint is imposed by limiting the coefficient of variation (COV) of indicated mean effective pressure (IMEP) throughout the test, as described in (D6). The same emission limits are imposed and the optimal control functions developed from the previous problem are employed as the initial estimates to initiate the procedure.

The driveability index (DI) for the previously optimized control is shown in Figure 12. The driveability index is the COV of imep normalized to the constraint limit. Although the emission constraints are satisfied and the fuel consumption is minimum, the driveability constraint is violated more than it is satisfied throughout the test.

The technique discussed earlier in the paper is employed to treat the driveability constraint, along with the previous constraints, and the driveability index after the final (i.e., the fifth) iteration is shown in Figure 13. Note that driveability is improved considerably, and the result of Figure 13 is considered to be acceptable. (There is little use in causing $C(t,u) \leq 0$ everywhere since the subjectiveness of the $C(t,u)$ - measure implies that it should be treated as a soft constraint, as opposed to a hard constraint such as an emission constraint.)

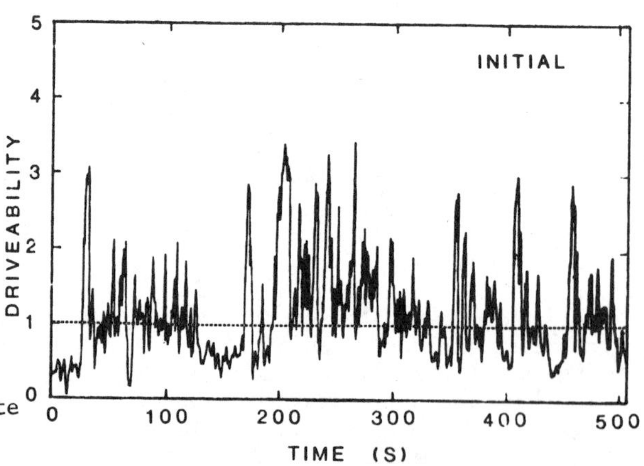

Fig. 12. Unconstrained Driveability

The results of the hot-start optimization with the driveability constraint are shown in Figure 14. The quantity labeled "driveability" in the figure is an L2-type norm of the driveability constraint violations, i.e.,

$$\text{Driveability} = \int_0^{t_f} \text{MAX}\ ((DI-1),0)^2 dt$$

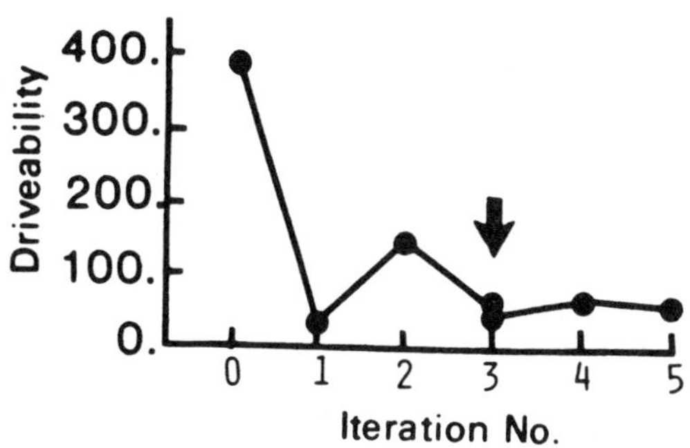

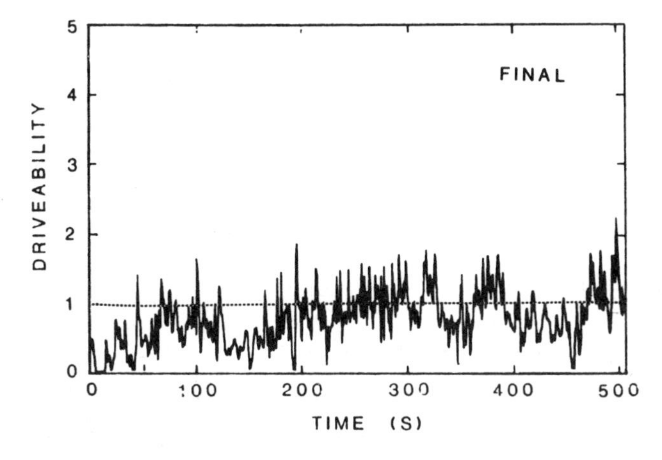

Fig. 13. Constrained Driveability

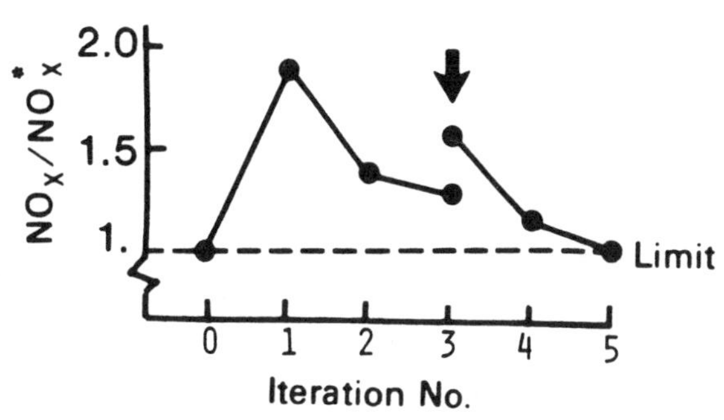

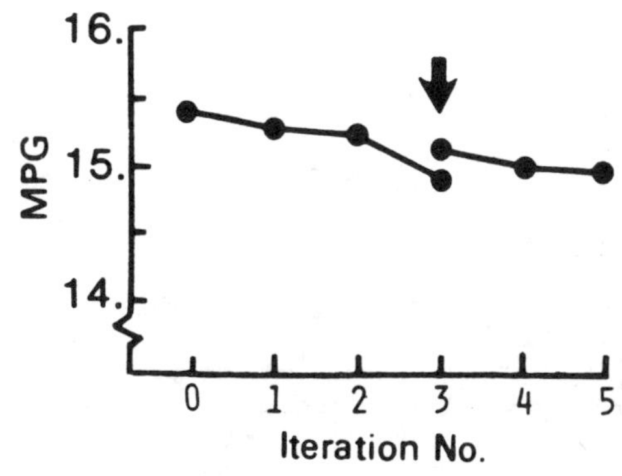

Fig. 14. Driveability-Constrained Fuel Economy & Emission Results

The larger the value, the worse the driveability. As can be seen from the figure, the initial control functions (Figure 7) satisfy the emission constraints and result in a fuel economy of 15.4 MPG. However, the resulting driveability is relatively poor. On the final iteration, the driveability is reduced by a factor of eight, the emission limits are satisfied, and the optimal fuel economy declines by only 0.4 MPG to 15.0 MPG.

One of the subtasks in imposing the driveability limit is the determination of the best type of computational algorithm to handle this type of constraint. An earlier section discusses both the stability and rate-of-convergence aspects of two different approaches:

1. the influence function approach, and

2. the interior-exterior penalty function approach.

The influence function approach is applied during the first and second iterations. Figure 14 shows that the first iteration dramatically improves the driveability but almost doubles the NO_x. The second iteration decreases the NO_x but increases the driveability norm. In both iterations, the fuel economy decreases. Although the influence function approach appears to be weighing the trade-off between emissions and driveability and moving in the appropriate direction, the erratic behavior of the drive-ability and NO_x is undesirable from a numerical stability standpoint.

The next three iterations are performed employing the interior-exterior penalty function approach. For these three iterations, the driveability norm, the emissions, and the fuel economy all converged smoothly, thus demonstrating relatively stable numerical behavior.

The optimal feedback control functions for air-fuel ratio and spark advance are shown in Figure 15. It is apparent that the driveability constraint causes a general enrichment of the air-fuel ratio and significant spark retard compared to the results shown in Figure 7. As shown in (D6), the COV of IMEP is particularly sensitive to air-fuel ratio. Consequently, it appears that driveability considerations force the enrighment of the air-fuel ratio. This requires more spark retard to control

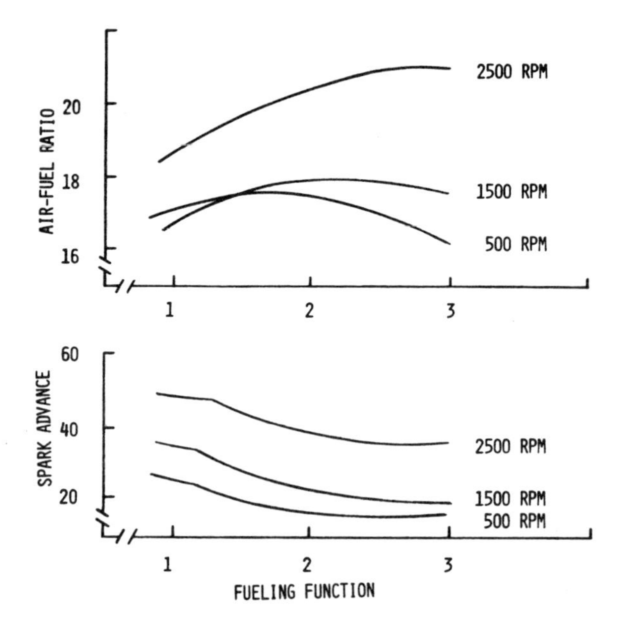

Fig. 15. Driveability Constrained Optimal Feedback Control Functions

the NO_x emission. However, the additional spark retard causes a decrease in engine efficiency and, in this case, a penalty of 0.4 MPG.

During the nominal run of the fourth iteration, a transmission downshift occurred on the second cycle acceleration to 55 MPH. This occurred because the transmission is vacuum modulated with a downshift scheduled for manifold and vacuum less than 1.5 inches of Hg. To avoid downshift, it is necessary to modify the controller, However, modifying the controller changes the system and, hence, the responses of the system as shown in Figure 14 by the discontinuities in the curves.

Although the system is changed, it is not necessary to restart the optimization from scratch but merely to continue iterating after the change is effected. This is a highly desirable characteristic not always available with optimization techniques based on engine mapping. For example, it would be catastrophic to a mapping technique if a certification engineer effected a change that required repeating a six-month program to redetermine the optimal engine calibration.

SUMMARY

This concludes the presentation of the three optimal automotive engine calibration problems. As is shown, the important considerations of the cold-start, driveability, catalytic converter dynamics, controller dynamics, and transient driving schedule are included in the analysis. The rapid convergence of this iteration procedur results from the advantageous utilization of the available knowledge about the process. This procedure is a powerful tool applicable in the automotive industry.

In this concluding paragraph the major results of this research are summarized and put into perspective with respect to possible current and future applications in the automotive industry. A gradient-based algorithm is presented for the solution of the optimal automotive engine control calibration problem. The mathematical model is replaced by the actual system because of the difficulties associated with modeling the transient system. The algorithm modifications necessitated by this substitution are discussed. The algorithm utilizes the advantageous formulation of the problem to minimize the required number of integrations of the system equations. This is important because the system equations are integrated experimentally, and this is a relatively expensive process. Unlike previous techniques, the gradient-type procedure presented herein considers the engine/vehicle system during transient operation, a surge-type driveability constraint, the cold-start portion of the FTP, and the catalytic converter transient emission signature.

REFERENCES

A1 Auiler, J. E., Zbrozek, J. D. and Blumberg, P.N. Optimization of Automotive Engine Calibration for Better Fuel Economy - Methods & Applications, SAE Paper No. 770076, February, 1977.

A2 Anderson, Robert L. Determination of Engine Cold Start Fuel Requirements Using a Programmable Electronic Choke, SAE Paper No. 770107, February, 1977.

A3 Amano, M., Naka, D.A. and Kobayashi, N. An Approach to Evaluate Vehicle Driveability Through Engine Dynamometer Testing, Third International Symposium on Automation of Engine and Emission Testing, September, 1974.

B1 Baker, R.E. and Daby, E.E. Engine Mapping Methodology, SAE Paper No. 77077, February, 1977.

B2 Bryson, A.E. and Ho, Y.C. Applied Optimal Control, Blaisdell Publishing Co., Waltham, Massachusettes, 1969, (Second Edition, Published by Halsted Press, 1975)

B3 Blumberg, P.N. Powertrain Simulation: A Tool for the Design and Evaluation of Engine Control Strategies in Vehicles, SAE Paper No. 760158, February, 1976.

B4 Borcherts, R.H., Stadler, H.L., Brehob, W.M., and Auiler, J.E. Improvements in Automotive Fuel Economy, American Association for the Advancement of Science, Energy Symposium, February, 1978.

C1 Cassidy, J.F. A Computerized On-line Approach to Calculating Optimum Engine Calibrations, SAE Paper No. 77078, February, 1977.

C2 Automotive Engineering, Controlling Emissions Without Catalysis, Volume 85 #7, July, 1977.

C3 Christensen, B.C. and Frank, A.A. The Fuel Saving Potential of Cars with Continuously Variables Transmissions and an Optimal Control Algorithm, ASME Publication No. 75-WA/Aut-20, December, 1975.

C4 Citron, S.J. Elements of Optimal Control, Holt Rinehart and Winston, Incorporated, New York, 1969.

C5 Cole, D. E. The Effect of Directed Mixture Motion on the Flame Kernel Development in a Constant Volume Bomb, ORA Project 05057, University of Michigan, May, 1966.

C6 Control of Air Pollution from New Motor Vehicles and New Motor Vehicle Engines, Federal Register, Volume 35, No. 219, Part II, November 10, 1970.

D1 Dohner, A.R. Transient System Optimization of an Experimental Engine Control System Over the Federal Emissions Driving Schedule, SAE Paper No. 780286, February, 1978.

D2 Daby, E., Brehol, W. and Baker, R. Engine Calibration Strategies for Emission Control: Part I Methodology, U.S. Department of Transportation Symposium, Cambridge, Massachusettes, July 8, 9, 1975.

D3 Draper, C.S., Li, Y.T. and Laning, H. Measurement and Control Systems for Engines, ASME Paper No. 49-SA-44, June, 1949.

D4 Draper, C.S. and LI, Y.T. Principles of Optimalizing Control Systems and an Application to the Internal Combustion Engine, ASME Publication, New York, September, 1951.

D5 Coordinating Research Council, Incorporated Driveability Instrumentation Tests, CRC Report No. 489, November, 1976.

D6 Dohner, A.R., Optimal Control Solution of the Automotive Emission-Constrained Minimum Fuel Problem, University Microfilm, August, 1978.

E1 Everett, R.L. Measuring Vehicle Driveability, SAE Paper No. 710137, January,1971.

E2 Engine Analyzer System, Tektronix Instruction Manual 070-0890-00, July, 1969.

F1 Fukushima, T., Nakamura, H. and Sakai, T. Exhaust Emission Control of S.I. Engines by Engine Modification - The SEEC System, SAE Paper 770224, February, 1977.

J1 Juneja, W.K., Horchler, D.D. and Haskew, H.M. A Treatise on Exhaust Emission Test Variability, SAE Paper No. 770136, February, 1977.

K1 Kuroda, H., Nakajuma, Y., Sugiharo, K., Takagi, Y. and Muranako, S. The Fast Burn with Heavy EGR, New Approach for Low NO_x and Improved Fuel Economy, SAE Paper No. 780006, February, 1978.

K2 Klimisch, R.L., Summers, J.C. and Schlatter, J. C. The Chemistry of Degradation in Automotive Emission Control Catalysts, GM Research Publication, GMR-1539, February, 1974.

L1 Lee, E.B. and Markus, L. Foundations of Optimal Control Theory, John Wiley & Sons Incorporated, 1967.

M1 McFarland, R.A. and Wood, C.D. An Analog Heat Release Computer for Engine Combustion Evaluation, SAE Paper No. 760553,June, 1976.

N1 Nakamura, H., Ohinouye, T., Hori, K., Kiyota, Y., Nakagansi, T., Akishino, K. and Tsukamoto, Y. Development of a New Combustion System (MCA-Jet) in Gasoline Engine, SAE Paper No. 780007, February, 1978.

N2 Nie, N.H., Hull, C.H., Jenkins, J.G., Steinbrenner, K. and Bent, D.H. Statistical Package for the Social Sciences, McGraw-

Hill Company, 1975.

P1 Prabhakar, R., Citron, S.J. and Goodson, R.E. Optimization of Automotive Engine Fuel Economy and Emissions, ASME Paper No. 75-WA/Aut-19, August, 1975.

P2 Patterson, D.J. Cylinder Pressure Variations, A Fundamental Combustion Problem. SAE Transactions Volume 75, 1967.

P3 Progress Report on Chrysler's Efforts to Meet the Federal Emission Standards for HC, CO and NO_x In 1979, 1980, 1981 and Subsequent Model Years, Chrylser Corporation, January, 1978.

P4 Paulsell, C.D. and Kruse, R.E. Test Varability of Emission and Fuel Economy Measurements Using the 1975 Federal Test Procedure, SAE Paper No. 741035, October,1976.

P5 Pontryagin, L.S. The Mathematical Theory of Optimal Processes, New York: Interscience, 1962.

R1 Rishavy, E.A., Hamilton, S.C., Ayers, J.A. and Keane, M.A. Engine Control Optimization for Best Fuel Economy with Emission Constraints, SAE Paper No. 770075, February,1977.

S1 Stivender, D.L. Development of a Fuel-Based Mass Emission Measurement Procedure, SAE Paper No. 710604, June, 1971.

S2 Simanaitis, D.J. Emission Test Cycles Around the World, Automotive Engineering, August, 1977.

S3 Stivender, D.L. Engine Air Control - Basis of a Vehicular Systems Control Hierarchy, SAE Paper No. 780346, February, 1978.

S4 Schweitzer, P.H., Volz, C. and DeLuca, F. Control System to Optimize Engine Power, SAE Paper No. 660022, January, 1966.

S5 Schweitzer, P.H., DeLuca, F. and Volz, C. Adaptive Control for Prime Movers, ASME Paper No. 67-WA/DGP-2, November, 1967

S6 Schweitzer, P.H. Control of Exhaust Pollution Through a Mixture Optimizer, SAE Paper No. 720254, January, 1972.

T1 Tanaka, M. and Durbin, E.J. Transient Response of a Carburetor Engine, SAE Paper No. 770046, February, 1977.

V1 Vora, L.S. Computerized Five Parameter Engine Mapping, SAE Paper No.77079, February, 1977.

W1 Waters, W.C. General Purpose Automotive Vehicle Performance and Economy Simulation, SAE Paper No.720043, January, 1972.

W2 Winsor, R.E. and Patterson, D.J. Mixture Turbulence-A Key to Cyclic Combustion Variation, SAE Paper No. 730086, January, 1973.

Z1 Zeilinger, K. and Hussman, A. The Influence of Transient Conditions on the Operation of an SI Engine, Especially with Respect to Exhaust Emissions, SAE Paper No. 750053, February 1975.

A COMPUTER-AIDED DESIGN OF ROBUST REGULATORS

J. S. Karmarkar* and D. D. Šiljak**

**Systems Control, Inc., Palo Alto, California, USA*
***University of Santa Clara, Santa Clara, California, USA*

Abstract. The objective of this paper is to present a design methodology for linear and nonlinear regulators by applying mathematical programming techniques. Starting with the linear vector equations

$$\dot{x}(t) = Px(t) + qu(t) \ , \ u(t) = -r^T x$$

an algorithmic procedure is outlined for selecting the regulator parameter vector r within the framework of classical time and frequency domain specifications, as well as the quadratic optimality criterion.

For nonlinear regulators described by nonlinear vector equations

$$\dot{x}(t) = Px(t) + q\phi[u(t)] \ , \ u(t) = -r^T x(t)$$

attention is focused on the design of absolutely stable regulators subject to prescribed exponential stability and sector maximization requirements.

Reformulating the performance specifications in terms of a set of inequalities, a feasible region is delineated in the regulator parameter space. Then, maximization of the volume of an imbedded hypercube (or hypersphere) inside the region via mathematical programming methods, results in an easily visualized solution set, this feature being particularly attractive in building robust regulators to meet real-world implementation tolerances and system parameter uncertainities.

I. INTRODUCTION

In order to realistically design increasingly complex control systems arising from modern technology subject to a myriad of performance specifications, it has become imperative that the computer plays a central role in a control system design. The major objective of this paper is to reinterpret the known results of control theory [1-5] in a mathematical programming format [6, 7] and formulate a comprehensive computer-aided design for both linear and nonlinear regulators. More precisely, the time and frequency domain specifications such as exponential stability, damping factor, steady-state error, time response peak, dominant root location and sensitivity, frequency response peak, integral-squared error; as well as the quadratic index optimality criterion (stated in terms of the return difference concept) for linear systems, and exponential stability and sector maximization requirements for absolutely stable nonlinear regulators, are used to form a set of inequalities, delineating a feasible region in the space of the design variables, which is called the regulator parameter vector space. Consequently, maximizing the volume of a hypercube (linear/nonlinear inequalities) or hypersphere (linear inequalities) imbedded in the feasible region completes the design - the specifications are satisfied for all regulator vectors lying within the imbedded volume.

It cannot be overemphasized that additional constraints on the parameters, arising for reasons of economy, realizability or availability, can be introduced into the design problem formulation in a natural manner, resulting in a more realistic design. Furthermore, although it is sometimes possible to obtain a specific set of numerical values for the parameters to satisfy design requirements, the practical designer is aware that neither are the design specifications accurate enough, nor is the data regarding the physical system precise; thus he is biased towards methods which provide a solution-set rather than a single solution point. It is for this reason that the approach proposed in [8, 9] and expanded in this paper, can be used to realistrically build robust regulators which can accommodate real world implementation tolerances and parameter uncertainties. It is not to say, however, that methods such as that of [10], which yield a unique numerical solution are not practically viable; useful information regarding the existence of a feasible set of parameters may be obtained by this technique. It is noteworthy that the methodology presented in this paper can also be used to obtain robustness with respect to sensor and actuator failures, as proposed in

[11], by the introduction of appropriate algebraic conditions. Finally, we should mention the fact that the methods and techniques developed here, can be used as an integral part of a comprehensive design of control schemes for interconnected large-scale systems [12], where the exponential stability constraint on the subsystem controllers is an essential requirement. By obeying the decentralized information structure of the overall system, the schemes can be used to design robust control capable of withstanding a broad class of perturbations in both feedback and subsystem interconnection structure.

Notation: With some obvious exceptions, Greek letters denote scalars, lower case Roman letters denote vectors, capital Roman letters denote matrices, and capital script letters denote sets.

2. IMBEDDING PROBLEM AND SOLUTION

The general mathematical programming problem [7] is to determine a vector $\hat{r}^* = (r_1^*, \ldots, r^*)^T$ that solves the problem of minimizing the objective function $f(\hat{r})$ subject to constraints; that is,

$$\text{minimize:} \quad f(\hat{r})$$
$$\text{subject to:} \quad h_i(\hat{r}) \geq 0, \; i = 1, 2, \ldots, \ell.$$
$$g_j(\hat{r}) = 0, \; j = 1, 2, \ldots, m.$$

where $\hat{r} \in R^n$ which is isomorphic to the Euclidean n-space; f, $\{h_i\}$, $\{g_j\}$ are continuous functions.

Consider the problem of imbedding a *hypersphere* in the feasible region $\mathcal{R}$, delineated by linear inequality constraints of the form,

$$a_{i0} + \sum_{j=1}^{n-1} a_{ij} r_j \geq 0, \; i = 1, 2, \ldots, \ell. \tag{2}$$

where a_{i0} and a_{ij} are constants. Normalizing the inequalities and introducing an auxiliary variable r_n (to represent the hypersphere radius) using the Chebyshev minimax formulation, the following mathematical programming problem [6] results:

$$\text{minimize:} \quad (-r_n)$$
$$\text{subject to:} \left\{ \left(a_{i0} + \sum_{j=1}^{n-1} a_{ij} r_j \right) \Big/ \left(\sum_{j=1}^{n-1} a_{ij}^2 \right)^{\frac{1}{2}} \right\} - r_n \geq 0, \; i = 1, q, \ldots, \ell \tag{3}$$

And (3) can be solved, using the well-known programming algorithm yielding $\hat{r}^* = (r^{*T}, r_n^*)^T$ where r_n^* represents the maximized hypersphere radius and $r^* = (r_1^*, r_2^*, \ldots, r_{n-1}^*)^T$ the hypersphere center, i.e., the solution set is:

$$\mathcal{R}^* \equiv \{r : \| r - r^* \| < r_n^*\}. \tag{4}$$

Moreover, the side of an inscribed hypercube is $2[r_n^*/(n-1)^{\frac{1}{2}}]$.

It is noteworthy that if $r_n^* \leq 0$, then one concludes that no feasible region $\mathcal{R}$ exists. In Section 4, it will be shown that the design of absolutely stable systems with a state variable feedback regulator can be cast into the format of equation (3).

Now, to generalize the above result to imbedding a of prescribed axial ratios and orientation, a transformation is called for such that (a) hyperellipses are transformed to hyperspheres of equal volume, (b) convexity of the transformed feasible region is retained. It sufficies to state that this can be accomplished by applying an appropriate equiaffine transformation [13], to the original set of variables [14].

Proceeding to consider the case occurring most often in practice, namely the inequalities are nonlinear. It is relevant to determine whether a feasible region exists. This is accomplished by solving the Chebyshev-like problem.

$$\text{minimize:} \quad (-r_n) \tag{5}$$
$$\text{subject to:} \; h_i(r) - r_n \geq 0, \; i = 1,2,\ldots,\ell.$$

where $r = (r_1, \ldots, r_{n-1})^T$; and the problem is infeasible if $r_n^* \leq 0$.

Provided the problem is feasible, a *hypercube* is imbedded in $\mathcal{R}$, using the formulation:

$$\text{minimize:} \quad (-r_n)$$
$$\text{subject to:} \; h_1(r_1 - ir_n, r_2 - jr_n, \ldots, r_{n-1} - kr_n) \geq 0$$
$$\vdots$$
$$h_\ell(r_1 - ir_n, r_2 - jr_n, \ldots, r_{n-1} - kr_n) \geq 0 \tag{6}$$

where $i, j, \ldots, k$ take on all possible combinations among 1, 0, -1. The corresponding solution set is:

$$\mathcal{R}^* \equiv \{r : |r_i - r_i^*| < r_n^*, \; i = 1,2,\ldots,(n-1)\}. \tag{7}$$

Note that the original ℓ inequalities are *each* decomposed into $3^{(n-1)}$ inequalities in (6). Although (6) appears formidable, it is really programmed by implementing (n-1) nested DO loops for *each* inequality as shown in [14] using the SUMT [7] computer program.

It is emphasized that the above formulation assumes that the region $\mathcal{R}$ delineated by the nonlinear inequalities is convex and therefore to imbed a hypercube, only the vertices of the hypercube need be tested. In practice, neither is it a trivial matter to test for convexity of a given nonlinear function, nor are the inequalities, arising from the simplest practical problems, convex. Consequently, an expedient strategy, employed in (6), is to check additional points lying on the hypercube surface and assume that if the inequality is not violated for these points, then it will not be violated in the interior of the hypercube. Moreover, parallelepiped of prescribed orientation and axial ratios can be imbedded, using the appropriate equi-affine transformation, described earlier.

3. LINEAR SYSTEM DESIGN

Consider a linear single input-output time-invariant system described by the vector differential equation

$$\dot{x}(t) = P\,x(t) + q\,u(t) \tag{8a}$$

$$u(t) = -r^T x(t) \tag{8b}$$

where (8a) represents the controlled object, and (8b) the regulator. P is a constant $N \times N$ matrix, q and r are constant N-vectors. And $x(t)$ is a real N-vector representing the state of the system. The corresponding open and closed loop characteristic equations are:

$$\Lambda(\lambda) \equiv \det(\lambda I-P) = \sum_{k=0}^{N} b_k(r)\lambda^k \tag{9}$$

$$\Gamma(\lambda) \equiv \det(\lambda I-P+qr^T) = \sum_{k=0}^{N} c_k(r)\lambda^k \tag{10}$$

where $\lambda = -\delta + i\omega$ represents the complex variable. It is well-known that the time and frequency domain performance of a linear system is essentially determined by the zeros of its characteristic polynomial (10). Our concern is to outline the manner in which some of the classical performance criteria can be expressed in terms of algebraic inequalities involving the regulator vector r.

Given (10), exponential stability and damping factor constraints can be expressed as a set of algebraic determinantal inequalities, as in [2, 14, 15]. Moreover, algebraic inequalities for dominant mode design, given the dominant pair locations and sensitivity can be developed using root-coefficient relations [14]. Furthermore, using the final value theorem of complex variable theory, a bound for the output steady state error, given a prespecified deterministic input can be formulated [14]. Finally, frequency and time response peak inequalities can also be developed,

The three basic design problems are:

Problem A - (Stability Design): Given (10), find the largest hypercube in regulator space, such that prescribed exponential stability/damping factor/steady state error specifications are satisfied; this entails formulating and solving the associated $n = (N+1)$ - dimensional mathematical programming problem (6).

Problem B - (Dominant Mode Design): Given (10), find the largest hypercube in regulator space, such that prescribed dominant root location and sensitivity specifications are met. In addition, the performance specifications of Problem A and time/frequency domain peak bounds can be superimposed [14]. Again this entails formulating and solving the associated $n = (N+1)$ - dimensional mathematical programming problem (6).

Enumerating the several advantages that accrue from the solution set format (7): Any r satisfying (7) is acceptable. If r^* is the operating point, all parameter variations within (7) are acceptable; r_n represents in quantative terms, the "stringency of the constraints" and nonfeasibility is indicated by $r_n^* \leq 0$. Moreover, note that as the dominancy constraint of Problem B is made increasingly stringent, the solution set (7) will be of decreasing volume; resulting in a trade-off between dominancy and allowable parameter variations. Finally, additional constraints, arising for any reason whatsoever can be introduced naturally into the problem formulation.

Besides the aforementioned classical performance specifications, it is advantageous to ensure that the system (8) is optimal with respect to some quadratic performance of the form:

$$J = \int_0^\infty (x^TQx + u^2)\,dt \tag{11}$$

where $Q = H^TH \geq 0$ is a nonnegative definite matrix. Then, under appropriate controllability and observability conditions (the rational functions $H^T(\lambda I-P)^{-1}q$ have no common cancellable factors), it can be shown [3, 16] that the stable control law $u(t)$ of (8b) is optimal with a prescribed degree of exponential stability $\delta \geq 0$, provided

$$\Pi(\omega^2, r, \delta) \equiv |\Gamma(-\delta+i\omega)|^2-|\Lambda(-\delta+i\omega)|^2 = \sum_{k=0}^{N} a_{2k}(r, \delta)\omega^{2k} \geq 0, \text{ for all } \omega \geq 0\,. \tag{12}$$

Although, given r, a numeric test is available to verify (13); for our purpose it is convenient to use the more conservative result [2, 3]:

A sufficient condition for (8) to be optimal with a prescribed degree of exponential stability $\delta \geq 0$ is that

$$a_{2k}(r, \delta) \geq 0,\ k = 0, 1, \ldots, N\,. \tag{13}$$

Clearly, (13) can be introduced into the framework of (6) naturally. Now, it is appropriate to present some justification for designing optimal regulators based on (13). In contrast to the Riccati equation approach, the conditions on the Q matrix in (11) have been relaxed, in that all possible nonnegative definite Q matrices are acceptable; this is valuable, since the relation between Q and the classical performance specifications is, at this time, somewhat obscure. On the other hand, it is known that systems optimal with respect to (11), exhibit several attractive properties. More specifically, from the classical viewpoint, the gain margin is infinite and the phase margin is at least 60°. Furthermore, the system displays tolerance to ($\frac{1}{2}$, ∞) sector nonlinearities, time delay, and sensitivity reduction with respect to parameter variations [5]. Thus our design philosophy is to retain these attractive features of the quadratic index, without numerically specifying Q and in addition impose the classical constraints:

Problem C - (Comprehensive Design): Given (8), find the largest hypercube in regulator space, such that prescribed optimality/exponential stability/damping factor/steady state error specifications are satisfied, i.e., solve associated n = (N+1)-dimensional mathematical programming problem (6).

Interestingly enough, if all states are not available, frequency domain suboptimality criteria [17] can be cast into this format.

In contrast to the aforementioned optimality constraint which does not consider specific input signals, it is sometimes relevant to require that the integral squared error (ISE) for a deterministic input be less than a prespecified value; this constraint is readily inserted in (6), using standard table of integrals [18]; similarly a squared error (MSE) bound for a statistical input signal with prespecified spectral characteristics can be imposed. Moreover, design situations may arise, wherein it is imperative that, rather than merely place an upper bound, MSE (say) be minimized subject to exponential stability and damping factor requirements [2]; clearly this class of problems is also ameanable to solution in the framework of the general mathematical programming problem (1). Finally, it is emphasized that although the present develepement used the state representation (8), the present method is applicable to any time-invariant system design, involving static (pure gain) or dynamic (lead/lag filters) compensators. Finally, the designer can also introduce the classical frequency domain criteria by reformulating the time and frequency response peak requirement, in terms of algebraic inequalities, as shown in the Appendix.

$$\dot{x} = \begin{bmatrix} 0 & 1 & 0 \\ 0 & -2 & -4 \\ 0 & 0 & -5 \end{bmatrix} x + \begin{bmatrix} 0 \\ 0 \\ -1 \end{bmatrix} u ,$$

which is to be compensated with state variable feedback $u = (r_1, r_2, r_3)x$.

The corresponding compensated and uncompensated characteristic equations are

$$\Lambda(\lambda) \equiv \sum_{k=0}^{3} b_k(r)\lambda^k \equiv \lambda^3 + 7\lambda + 10 = 0$$

$$\Gamma(\lambda) \equiv \sum_{k=0}^{3} c_k(r)\lambda^k \equiv \lambda^3 + (7+10r_2+10r_3)\lambda^2$$

$$+ (10 + 10r_1 + 10r_2 + 20r_3) + 10r_1 = 0 \quad (15)$$

and we require the system to be optimal [3], the degree of exponential stability $|\delta| > 1$ and parameter constraints $|r_i| > 20$, i = 1, 2, 3 ; resulting in the inequalities:

optimality constraints

$$r_1 > 0$$

$$100(r_2+r_3)+100r_3+120r_2 - 20r_1 > 0$$

$$100(r_1^2+r_2+4r_3^2) + 200(2r_2r_3+r_1r_3) + 20(3r_1+10r_2+20r_3) > 0 \quad (16)$$

exponential stability constraints

$$-4 + 10r_3 > 0$$

$$4 + 10r_2 + 10r_3 > 0$$

$$10r_1 - 10r_2 - 1 > 0$$

$$(4+10r_2+10r_3)(10r_1-10r_2-1)+(4+10r_3) > 0 \quad (17)$$

parameter constraints

$$20 - |r_i| > 0, \; i = 1, 2, 3 \quad (18)$$

relative stability constraints

$$c_i(r) > 0, \; \Delta_i(r) > 0, \; i = 2, 3, 4, 5 \quad (19)$$

$$\Delta_5(r) = \begin{vmatrix} 4\zeta^2-1 & -2\zeta(1-\zeta^2)^{\frac{1}{2}}c_3(r) & (1-\zeta^2)^{\frac{1}{2}}c_2(r) & 0 & 0 \\ -\zeta(4\zeta^2-3) & (2\zeta^2-1)c_3(r) & -\zeta c_2(r) & 0 & 0 \\ 0 & 4\zeta^2-1 & -2\zeta(1-\zeta^2)^{\frac{1}{2}}c_3(r) & (1-\zeta^2)^{\frac{1}{2}}c_2(r) & 0 \\ 0 & 4\zeta^2-3 & (2\zeta^2-1)c_3(r) & -\zeta c_2(r) & c_1(r) \\ 0 & 0 & 4\zeta^2-1 & -2\zeta(1-\zeta^2)^{\frac{1}{2}}c_3(r) & (1-\zeta^2)c_2(r) \end{vmatrix} \quad (20)$$

As an example of Problem C, consider a third order servomechanism

Using the starting vector $(1.2, 1, -0.5, 0.01)^T$ the results obtained on an IBM 360/67 are summarized as follows:

ζ	r_1^*	r_2^*	r_3^*	r_4^*	Execution Time (Sec)
0.250	16.95	10.68	-3.44	3.04	538
0.500	16.95	10.68	-3.44	3.04	505
0.707	16.95	10.77	-3.44	2.95	400

Table 1. Effect of damping factor ζ on r^*

It is noted that the solution sets given by (7) for $\zeta = 0.25$ and 0.5 are identical implying the constraints (19) are not stringent for these values. This is to be expected since an optimal system approximates a Butterworth type pattern, which for a third order system has complex poles with $\zeta = 0.5$. We should also mention that the determinants in (19) are calculated on the machine, and the problem is not quite as formidable as it appears.

4. NONLINEAR SYSTEM DESIGN

Let us consider the Lur'e-Postnikov system

$$\dot{x}(t) = Px(t) + q\phi[u(t)] \quad (21a)$$

$$u(t) = -r^T x(t) \quad (21b)$$

where the matrix P is Hurwitz; the nonlinearity $\phi(u)$ is a real valued, continuous function of a real scalar u, and belongs to the class Φ_κ: $\phi(0) = 0$, $0 \leq u\phi(u) \leq \kappa u^2$, $\kappa < +\infty$.

Provided the solution $x(t, r)$ of (21) is well defined for each r of the parameter space, then system (21) is said to be absolutely stable if the equilibrium $x = 0$ is globally asymptotically stable for any $\phi(u) \in \Phi_\kappa$ and $r \in R$. [2, 3].

Assuming (21) to be completely controllable and observable, it can be shown [2, 3] that the system is exponentially absolutely stable provided for some real θ

$$\Pi(\omega^2, r, \kappa, \theta, \delta) \equiv |\det(i\omega I - P_\delta)|^2 \Big[\kappa^{-1} + \mathrm{Re}\{(1+i\omega\theta)\, r^T(i\omega I - P_\delta)^{-1} q\} - \delta\theta\kappa |r^T(i\omega I - P_\delta)^{-1} q|^2\Big] > 0, \text{ for all } \omega \geq 0 \quad (22)$$

and $P_\delta = P + \delta I$ is Hurwitz, where $\delta \geq 0$, is the prescribed degree of exponential stability. Noting that (22) is an even polynomial we have

$$\Pi(\omega^2, r, \kappa, \theta, \delta) \equiv \sum_{k=0}^{N} b_{2k}\omega^{2k} \geq 0, \quad \text{for all } \omega \geq 0 \quad (23)$$

where the coefficients $b_{2k} = b_{2k}(r, \kappa, \theta, \delta)$ are real and $b_{2N} \neq 0$. And consequently, we obtain the following (somewhat conservative) result [2]:

A sufficient condition for (21) to be exponentially absolutely stable is that

$$b_{2k}(r, \kappa, \theta, \delta) \geq 0, \quad k = 0, 1, \ldots, N. \quad (24)$$

It is emphasized that the necessary and sufficient conditions for absolute stability are not known at the present time; thus R cannot be determined. But an estimate of this region can be obtained using the sufficient conditions (24). The following well-known design formulations result from practical considerations:

<u>Problem D</u>: Given (21), find the regulator vector $r^* \in R$, such that the setor $[0, \kappa]$ of the class Φ_κ is maximized. Clearly, this can be solved by the nonlinear program, in (N+2) dimension:

$$\text{minimize:} \quad (\kappa^{-1})$$

$$\text{subject to: } b_{2k}(r, \kappa, \delta, \theta) \geq 0, \; k = 0, 1, \ldots, N \quad (25)$$

<u>Problem E</u>: Given (21) find the largest region $R^* \subset R$, such that for all $r \in R^*$ the system is exponentially absolutely stable, in a specified sector κ_1; the prescribed degree of exponential stability being δ.

It is of particular interest to note that for state variable feedback regulators, i.e., for (21b), the inequalities (24) are linear, provided θ is specified; therefore the hypersphere imbedding scheme and the associated n=(N+1)-dimensional linear programming formulation (3) is applicable. In general, for each θ_i, we obtain a corresponding $R^*_{\theta_i}$ and R^* is the union of these regions, i.e.

$$R^* = \cup R^*_{\theta_i} \quad (26)$$

where

$$R^*_\theta = \{r : \theta = \theta_i, \; \kappa = \kappa_1 : ||r - r^*|| < r_n^*\}. \quad (27)$$

Finally, it is possible to combine Problems D and E to obtain:

<u>Problem F</u>: Given (21), find the largest region $R^* \subset R$ such that for all $r \in R^*$ and all $\phi(u) \in \Phi_{\kappa_m}$ the maximized class, the system is exponentially absolutely stable.

It cannot be overemphasized that, from a mathematical programming viewpoint, the partitioning of parameters into θ, κ, δ, r is artificial. Moreover, in practice, the partioning of (21) into the controlled object and regulator equations is often invalid (for example, when dynamic compensating lead and lag filters are used). Finally, although the regulator equation in this and the preceding sections was restricted to the feedback gain form, in conformity with current literature, the method is by no means restricted to this regulator equation.

As an illustration of Problems D, E, and F formulated above, consider a unity feedback system of the Lur'e-Postnikov class with transfer function of the linear part:

$$G(\lambda, r) = \frac{\lambda^2 + r_2\lambda + r_3}{r_1(\lambda+1)(\lambda+2)(\lambda+3)} . \tag{28}$$

For $\theta = 0$ (time-varying nonlinearity in the sector $[0, r_1^{-1}]$) we get [2],

$$\pi(\omega^2, r) \equiv r_1\omega^6 + (14r_1 - r_2 + 6)\omega^4 + (49r_1 + 11r_2 - 6r_3 - 6)\omega^2 + 6(6r_1 + r_3) . \tag{29}$$

From (24), we obtain

$$\begin{aligned} r_1 &> 0 \\ 14r_1 - r_2 + 6 &\geq 0 \\ 49r_1 + 11r_2 - 6r_3 - 6 &\geq 0 \\ 6r_1 + r_3 &> 0 \end{aligned} \tag{30}$$

Then, Problem D is to minimize r_1 subject to (30). Although this problem is trivial in that for vanishingly small r_1 the constraints are satisfied in a triangular region of the r_2r_3 plane, it is not always possible to make such astute observations and a computer role is essential. In this case, however, we obtain

$$\begin{aligned} 6 - r_2 &\geq 0 \\ 11r_2 - 6r_3 - 6 &\geq 0 \\ r_3 &\geq 0 . \end{aligned} \tag{31}$$

We determine the solution set for Problem F by imbedding a circle in the feasible region delineated by (31). Utilizing (3), we get:

Maximize: $(-r_4)$

subject to:

$$\begin{aligned} 6 - r_2 - r_4 &\geq 0 \\ 11r_2 - 6r_3 - 6 - \sqrt{157}\, r_4 &\geq 0 \\ r_3 - r_4 &\geq 0 \end{aligned} \tag{32}$$

Starting with the initial vector $(3, 2.5, 0.01)^T$, the execution time on an IBM 360/67 was 0.99 sec, yielding the solution set for Problem F:

$$R^* = \{(r_2, r_3)^T : \theta = 0,\ 0 < \kappa < +\infty ,\ ||r - (3.97, 2.03)^T|| < 2.03\} . \tag{33}$$

Now, for $\theta \neq 0$, from (28) we get

$$\pi(\omega^2, \theta, r) \equiv (\theta + r_1)\omega^6 + [14r_1 + (6\theta - 1)r_2 - \theta r_3 + (6 - 11\theta)]\omega^4 + [49r_1 + (-6\theta + 11)r_2 + (11\theta - 6)r_3 - 6]\omega^2 + 6(6r_1 + r_3) . \tag{34}$$

For prescribed values of θ, spheres are imbedded in the three-dimensional $r_1r_2r_3$ space subject to (24) and the parameter constraint $r_1 \leq 10$. The results are summarized as follows:

θ	r_1^*	r_2^*	r_3^*	r_4^*	EXECU. TIME (SEC)
-2	6.000	-3.948	-6.329	4.000	1.81
-1	5.500	-1.977	-2.041	4.499	1.83
0	5.000	4.377	3.062	4.999	1.56
1	4.456	8.780×10^{13}	1.400×10^{14}	5.329	2.34
2	3.701	7.180×10^{13}	2.600×10^{14}	5.269	2.39

Table 2. Effect of parameter θ on r^*

which verifies the claim that the regions in (26) and (27) can be evaluated numerically in a relatively efficient way. The computation procedure can be extended to include computations of finite stability regions in the state space when the nonlinear characteristics violate the sector constraints. This is described in reference [9].

5. COMPUTER ALGORITHM

Keeping in mind the objective of automated design procedures, namely, to free the designer of all tasks, except the qualitative decision making phase; a computer algorithm is outlined in this section. The design cycle, described in the sequel, depicts the manner in which the designer "converges" to an "optimum" design by successive iteration on the design parameters. The various phases of this cycle are considered in some detail as shown in Figure 1.

The *control problem* formulation phase involves data input to the computer specifying the system configuration, the regulator configuration and the performance specifications to be satisfied.

Based on this information, the computer proceeds to the corresponding *programming problem* formulation phase, wherein using symbol manipulation algorithms and well-known signal flowgraph methods [19], the set of inequalities resulting from the performance specifications are generated and the imbedding problem is set up.

subsequently, the computer proceeds to the *imbedding problem* solution phase, which entails imbedding a hypercube (hypersphere) in the feasible region, using for example, the SUMT [7] program shown in Figure 2. The essence of the SUMT algorithm is to transform the constrained minimization problem (1) to a sequence of unconstrained minimization problems by forming the penalty function $P(r, \varepsilon)$ shown in Figure 2. Then, it can be shown [7] that as $\varepsilon \to 0$ the sequence of minima generated by minimizing $P(r, \varepsilon)$ converges to the solution of the constrained minimization problem (1).

Having obtained numerical values for the imbedding problem, *simulation* of the system for trial points within the hypercube (hypersphere) is automatically initiated; using for example, the 360 CSMP program and the results are displayed on a CRT terminal (ideally) or a plotter.

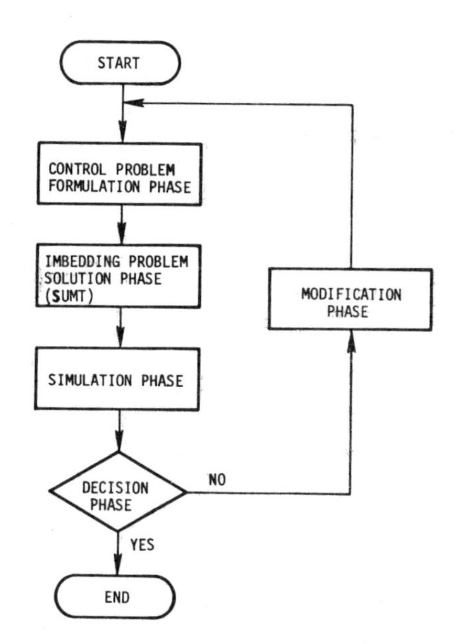

Figure 1. Computer-Aided Design Phases

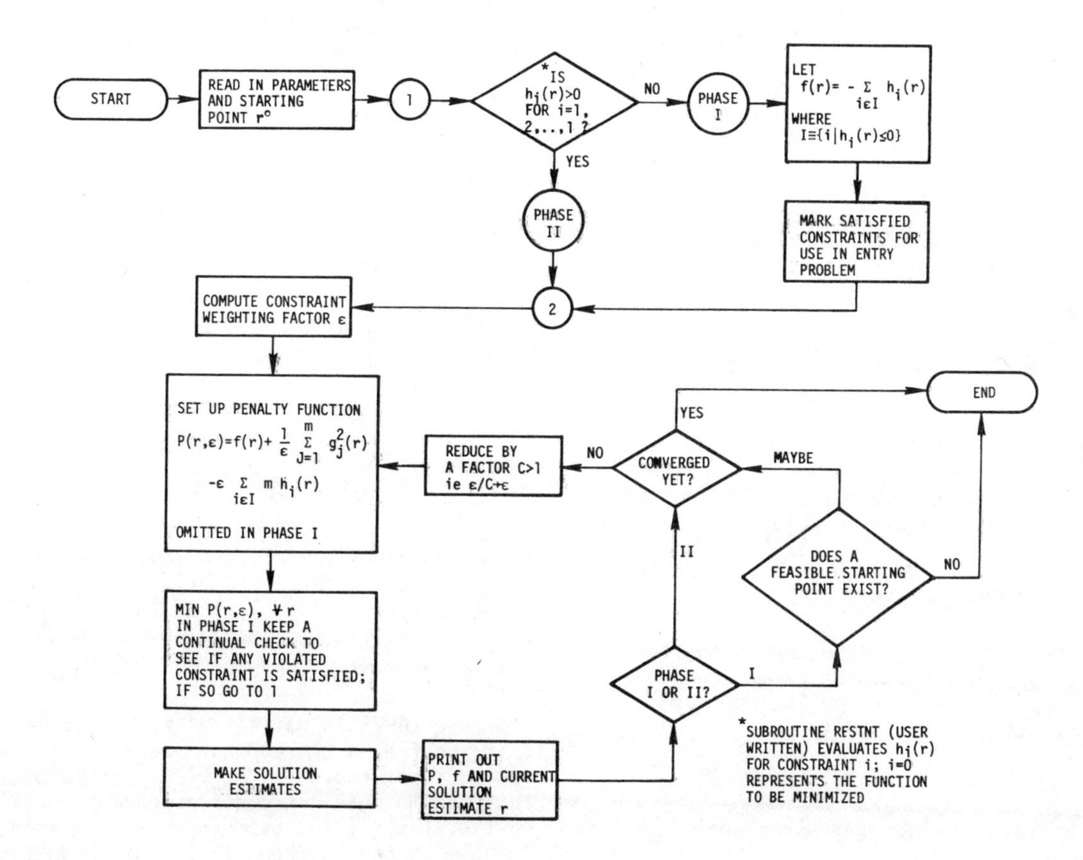

Figure 2. Overview of SUMT

At this stage, the computer enters the *decision* phase and program control is transferred to the keyboard, manned by the designer. The designer is now called upon to make a qualitative decision as to whether the current stimulation results are satisfactory; if they are not, the designer initiates the *modification* phase by appropriate alterations in the regulator configuration and performance specifications. His choice being guided by the results of the imbedding, simulation programs and experience.

At the present time, such a comprehensive

design package is not available and further work in this direction remains to be done. More specifically, a subprogram for the programming problem formulation phase, using well-known signal flowgraph methods [19], needs to be implemented. Moreover, while the control problem, decision and modification phases, although presently unavailable, do not constitute a major programming effort; the linking of the various phases into a compatible package does not pose difficulties. For the purposes of the present paper, the first two phases were implemented by hand and a straightforward modification of the SUMT [7] program, allowed direct utilization of this program for the imbedding solution phase. It is of interest to note that the SUMT computer code typically required 80 K bytes of memory and total run time (compile/link/schedule/execute) of the order of 15 seconds on a CDC 6400; while on an IBM 360/67 using double precision, the former was 180 K and the latter approximately 60 seconds.

6. CONCLUSION

A unified computer-oriented method was proposed for the design of linear time-invariant regulators, for linear and nonlinear systems, using mathematical programming methods.

A unique feature of the formulation was that it allowed one to simultaneously consider classical and modern performance specifications. It was noted that practical constraints arising for reasons of economy, availability and physical realizability could be naturally introduced into the mathematical programming format, resulting in a more realistic design. Moreover, since one obtained a solution-set, rather than a solution-point, practical implementation did not entail accurate parameter settings.

Further work remains to be done in implementing the automated design procedure outlined in Section 5 and subsequently extending the program to handle multivariable systems using known results [20, 21].

ACKNOWLEDGEMENTS

The research reported herein was supported in part by the U.S. Department of Energy, Division of Electric Energy Systems, under the Contract EC 77-S-03-1493.

APPENDIX

FREQUENCY/TIME DOMAIN RESPONSE BOUNDS

It is well-known to designers using classical methods that control bandwidth is an important design specification, often overlooked by optimal control methods. Also, the effect of numerator dynamics is not adequately treated by the various analytical methods. To alleviate this situation, algebraic inequalities providing a bound on time and frequency response peaks and bandwidth are developed.

The magnitude of the frequency response curve can be expressed by $M(\omega, r)$ where

$$| M(\omega, r) |^2 = T(\lambda)\, T(-\lambda) \Big|_{\lambda=i\omega} \qquad \text{(A-1)}$$

and

$$T(\lambda) = H^T(\lambda I - P - qr^T)^{-1}$$

$$= \frac{N_e(\lambda, r) + N_o(\lambda, r)}{D_e(\lambda, r) + D_o(\lambda, r)} \Bigg|_{\lambda=i\omega} , \qquad \text{(A-2)}$$

where the subscript e and o represent the even and odd part respectively. From (A-1) and (A-2) we obtain

$$| M(\omega, r) |^2 = \frac{N_e^2 - N_o^2}{D_e^2 - D_o^2} = \frac{\sum_{k=0}^{K} A_k(r)\, \omega^{2k}}{\sum_{k=0}^{N} B_k(r)\, \omega^{2k}} . \qquad \text{(A-3)}$$

Now by placing a bound M_p^2 on the frequency response magnitude curve (A-3), "good performance" can be expected. Thus

$$M_p^2 \geq |M(\omega, r)|^2 , \text{ for all } \omega \geq 0 . \qquad \text{(A-4)}$$

Consequently

$$M_p^2 \sum_{k=0}^{N} B_k(r)\omega^{2k} - \sum_{k=0}^{K} A_k(r)\omega^{2k} \geq 0 ,$$

$$\text{for all } \omega \geq 0 . \qquad \text{(A-5)}$$

Since $N \geq K$ for control systems (A-5) is rewritten as

$$\sum_{k=0}^{N} C_k(M_p, r)\omega^{2k} \geq 0, \text{ for all } \omega \geq 0. \qquad \text{(A-6)}$$

For a numerically specified regulator vector r , (A-6) can be verified using the nonnegativity test described in [22, 23]. Alternatively, a sufficient condition for (A-6) is

$$C_k(M_p, r) \geq 0, \; k = 0, 1, \ldots , N . \qquad \text{(A-7)}$$

Provided (A-7) is satisfied, we are assured that the frequency response magnitude is bounded from above by M_p . Now it is of interest to obtain the corresponding bound on the time domain response to a unit step input. It can be shown [24] that, when the magnitude curve has a single maximum M_{max} , the output response peak $y(t)$ is bounded by the relations

$$y(t)_{max} \leq 1.18\, M_{max} \leq 1.18\, M_p . \qquad \text{(A-8)}$$

In addition to responding to the input signal, a control system must be able to reject noise and unwanted signals. Large M_{max}

results in a sharper cutoff characteristic; but this must be traded off against a large overshoot, indicated by (A-8). Moreover, the noise rejection of filtering characteristics are roughly reflected by the bandwidth. Thus, it is of interest to formulate algebraic inequalities which guarantee that the (3db down) bandwidth ω_b is bounded by the relation

$$\omega_1 < \omega_b < \omega_2 \quad . \tag{A-9}$$

This can be expressed as

$$|M(\omega, r)|^2 \geq \frac{1}{2}\left(\frac{A_0(r)}{B_0(r)}\right)$$

$$\text{for all } \omega:\ 0 < \omega < \omega_1 \tag{A-10}$$

and

$$|M(\omega, r)|^2 \geq \frac{1}{2}\left(\frac{A_0(r)}{B_0(r)}\right)$$

$$\text{for all } \omega:\ \omega_2 < \omega < \infty \ , \tag{A-11}$$

where $|M(\omega, r)|$ is given by (A-3). These equations can be rearranged to yield the polynomial inequalities

$$F_1(\alpha, r) \equiv |M(\frac{1}{\alpha} + \frac{1}{\omega_1}\ , r)|^2$$

$$- \frac{1}{2}\left(\frac{A_0(r)}{B_o(r)}\right) \geq 0 \ , \text{ for all } \alpha \geq 0 \tag{A-12}$$

$$F_2(\beta, r) \equiv \frac{1}{2}\left(\frac{A_0(r)}{B_0(r)}\right) - |M(\beta-\omega_2, r|^2 \geq 0 \ ,$$

$$\text{for all } \beta \geq 0 \ . \tag{A-13}$$

A sufficient condition to satisfy these inequalities is that all the coefficients of the polynomials $F_1(\alpha, r)$ and $F_2(\beta, r)$ be nonnegative. Necessary and sufficient conditions using the modified Routh algorithm are available elsewhere [2].

To summarize, algebraic inequalities have been developed to enable the designer to place bounds on the time and frequency response peaks (A-7) and (A-8), as well as the bandwidth (A-12) and (A-13).

REFERENCES

[1] Thaler, G.J., and R.G. Brown, Analysis and Design of Feedback Control Systems, McGraw-Hill, New York, 1960.

[2] Šiljak, D.D., Nonlinear Systems: Parameter Analysis and Design, Wiley, New York, 1969.

[3] Šiljak, D.D., Algebraic criterion for absolute Stability, optimality, and passivity of dynamic systems, Proceedings of IEE, 117(1970), 2033-2036.

[4] Karmarkar, J.S., Multiparameter design of linear optimal regulators with prescribed degree of exponential stability, Proceedings of the Fourth Asilomar Conference on Circuits and Systems, Pacific Grove, California, 1970, pp. 161-164.

[5] Anderson, B.D.O., and J. B. Moore, Linear Optimal Control, Prentice-Hall, Englewood Cliffs, New Jersey, 1971.

[6] Zukhovitskii, S.I., and L.I. Avdeyeva, Linear and Convext Programming, Saunders, New York, 1966.

[7] Fiacco, A.V., and G.P. McCormick, Nonlinear Programming: Sequential Unconstrained Minimization Technique, Wiley, New York, 1968.

[8] Karmarkar, J.S., and D.D. Šiljak, A computer-aided regulator design, Proceedings of the Ninth Annual Allerton Conference on Circuits and Systems, University of Illinois, Monticello, Illinois, 1971, pp. 585-594.

[9] Karmarkar, J.S., and D.D. Šiljak, Maximization of absolute stability regions by mathematical programming methods, Regelungtechnik, 2 (1975), 59-61.

[10] Zakian, V., and U. Al-Naib, Design of dynamical and control systems by the method of inequalities, Proceedings of IEE, 120 (1973), 1421-1427.

[11] Ackerman, J.E., A robust control system design, Proceedings of the 1979 Joint Automatic Control Conference, Denver, Colorado, (to appear).

[12] Šiljak, D.D., Large-Scale Dynamic Systems: Stability and Structures, North-Holland, New York, 1978.

[13] Eisenhart, L.P., Coordinate Geometry, Dover, New York, 1960.

[14] Karmarkar, J.S., A Regulator Design by Mathematical Programming Methods, Ph.D. Thesis, University of Santa Clara, Santa Clara, California, 1970.

[15] Stojić, M.R., and D.D. Šiljak, Generalization of the Hurwitz, Nyquist, and Mikhailov stability criteria, IEEE Transactions, AC-10(1965), 250-255.

[16] Kalman, R.E., When is a linear control system optimal?, ASME Transactions, 86 (1964), 51-60.

[17] Canales, R., A lower bound on the performance of optimal regulators, IEEE Transactions, AC-15(1970), 409-415.

[18] Newton, G.C., L.A. Gould, and J.F. Kaiser, Analytic Design of Linear Feedback Control, Wiley, New York, 1957.

[19] Dunn, W.R., and S.P. Chan, Flowgraph analysis of linear systems using remote time-shared computation, Journal of the Franklin Institute, 288(1969), 337-349.

[20] Šiljak, D.D., New algebraic criteria for positive realness, Journal of the Franklin Institute, 290(1971), 109-120.

[21] Šiljak, D.D., A criterion for nonnegativity of polynomial matrices with application to system theory, Publications de la faculte d'electrotechnique, de l'universite a Belgrade, 79-96(1973), 163-172.

[22] Karmarkar, J.S., On Šiljak's absolute stability test, Proceedings of IEEE, 58(1970), 817-819.

[23] Jury, E.I., Inners and Stability of Dynamic Systems, Wiley, New York, 1974.

[24] Solodovnikov, V.V., Introduction to Statistical Dynamics of Automatic Control Systems, Dover, New York, 1960.

NONLINEAR PROGRAMMING FOR SYSTEM IDENTIFICATION

N. K. Gupta

Systems Control, Inc. (Vt), 1801 Page Mill Road, Palo Alto, CA 94304, USA

Abstract. Numerical procedures for dynamic system identification are discussed. Efficient algorithms for static least-squares problems provide a starting point for dynamic systems nonlinear programming methods. This paper shows that in dynamic systems, the additional computation time required for the first and the second gradients over function evaluation is small compared to static systems. This makes gradient procedures very attractive for dynamic system parameter estimation. Additional simplifications are made for linear systems. Finally, some practical simplifications are suggested to enable identification in large scale systems using current computers.

Keywords. System identification, nonlinear programming, parameter estimation, numerical procedures, dynamic systems, maximum likelihood method.

I. LEAST-SQUARES PARAMETER ESTIMATION

Systematic methods for parameter estimation from imprecise measurements can be traced back to Gauss. Since Gauss's least-square estimation formulation possesses attractive properties, much attention has been given to efficient computer solutions of the resulting optimization problem. State-of-the-art improvements have been achieved due to Golub [1], Gill and Murry [2], Bierman [3], Nazareth [4], Dennis [5], Golub and Pereyra [6], and others.

Consider a linear estimation problem where p parameters θ are related to an output variable y_i according to the following equation:

$$y_i = x_i^T\theta + \varepsilon_i \qquad i = 1, 2, \ldots, N \tag{1}$$

Here y_i and px1 vector x_i are measured and ε_i represents the total error. The least-squares estimates are based on minimizing

$$J(\theta) = \frac{1}{2}\sum_{i=1}^{N}(y_i - x_i^T\theta)^2 \tag{2}$$

A direct differentiation of the above equation with respect to θ gives the estimate $\hat{\theta}$.

$$\hat{\theta} = (X^TX)^{-1}X^TY \tag{3}$$

where X is the N×p matrix of x_i^T and Y is the N×1 column vector of y_i. Though Equation (3) is an explicit solution to the problem of parameter estimation, it is also numerically sensitive and the computational requirements are unnecessarily high.

To improve numerical conditioning, we must start with the system of equations

$$Y = X^T\theta \tag{4}$$

and convert it into the following form by orthogonal transformations (An orthogonal transformation preserves the length of the vector)

$$\begin{matrix} p\,\{ \\ N-p\,\{ \end{matrix} \underbrace{\begin{bmatrix} Z \\ E \end{bmatrix}}_{1} = \underbrace{\begin{bmatrix} U \\ 0 \end{bmatrix}}_{p}\theta \tag{5}$$

The parameter estimates are then obtained by solving (using backward substitution)

$$U\theta = Z \tag{6}$$

and E^TE is the residual mean square value.

A new measurement may be added if U and Z are saved. We start with

$$\begin{bmatrix} Z \\ y_{N+1} \end{bmatrix} = \begin{bmatrix} U \\ x_{N+1}^T \end{bmatrix}\theta \tag{7}$$

and reduce it to a new upper triangular form.

$$\begin{bmatrix} Z_{N+1} \\ e_{N+1} \end{bmatrix} = \begin{bmatrix} U_{N+1} \\ 0 \end{bmatrix}\theta \tag{8}$$

A new estimate is obtained directly without explicit need for previous measurements. Estimates may also be obtained if one or more components of θ are dropped from the estimated set. The approach outlined above is closely related to the square-root filtering approach of Kaminski and Bryson [7] and Bierman [3].

An important modification to the least-square formulation is the method of ridge regression [17]. Ridge regression uses a modified performance index

$$J_r(\theta) = \sum_{i=1}^{N} (y_i - x_i^T\theta)^2 + \lambda \sum_{i=1}^{p} w_i(\theta_i - \theta_i^o)^2 \qquad (9)$$

This performance index reflects a priori knowledge of the parameters θ_i^o. The weight factors w_i indicate relative confidence in the postulated values θ_i^o. In the analysis of a problem, θ estimates for several λ values may be computed.

II. NONLINEAR ESTIMATION PROBLEM

In many systems, the measurement y_i is a nonlinear function of θ,

$$y_i = f(x_i^T,\theta) + \varepsilon_i \qquad (10)$$

The least-squares estimate is obtained by minimizing the following criterion:

$$J(\theta) = \frac{1}{2}\sum_{i=1}^{N}\left(Y_i - f(x_i^T,\theta)\right)^2 \qquad (11)$$

An explicit solution cannot be written for the θ estimate. An iterative solution must therefore be adopted in the absence of a direct solution. The first and the second gradients of $J(\theta)$ are

$$\frac{\partial J}{\partial \theta} = - X_k^T \Delta Y_k \qquad (12)$$

$$\frac{\partial^2 J}{\partial \theta^2} = X_k^T X_k + \Delta M_k \qquad (13)$$

The addition term ΔM in the second gradient matrix is of the form

$$(\Delta M_k)_{\ell m} = \sum_{i=1}^{N} \frac{\partial^2 f(x_i^T,\theta_k}{\partial\theta_\ell \partial\theta_m} \left(y_i - f(x_i^T,\theta_k)\right) \qquad (14)$$

The following iterative solution procedures may be used to minimize $J(\theta)$

Step 1: Choose an initial guess for θ, say θ_o.

Step 2: Compute $\left.\frac{\partial J}{\partial \theta}\right|_{\theta_k}$, $\left.\frac{\partial^2 J}{\partial \theta^2}\right|_{\theta_k}$

Step 3: Solve the following equation for $\Delta\theta$.

$$\frac{\partial^2 J}{\partial \theta^2}\Delta\theta = - \frac{\partial J}{\partial \theta} \qquad (15)$$

Numerically $\Delta\theta$ is obtained from a set of normal equations.

Step 4: Set $\theta_{k+1} = \theta_k + \Delta\theta$

Step 5: Test for convergence. For example, is $||\Delta\theta||$ less than some limit? If not, go to Step 2.

Nonlinear least-squares includes linear least-squares as one step. In addition, the derivatives of f with respect to θ must be computed in each iteration. In general, quadratic convergence is obtained if ΔM is included in the second gradient computation.

Computation of Gradients of f. The gradients of f may be determined analytically or numerically. Often the numerical difference approximation is faster and more accurate. The first gradient in direction ξ_i in the parameter space at θ is obtained by the approximation

$$\frac{\partial f(\theta)}{\partial \xi_i} = \frac{f(\theta + \Delta_i\xi_i) - f(\theta)}{\Delta_i} \qquad (16)$$

or by a better two-sided approximation

$$\frac{\partial f(\theta)}{\partial \xi_i} = \frac{f(\theta + \Delta_i\xi_i) - f(\theta - \Delta_i\xi_i)}{2\Delta_i} \qquad (17)$$

where Δ_i is a small fraction. The second gradient in directions ξ_i and ξ_j is obtained by the approximation

$$\frac{\partial^2 f(\theta)}{\partial\xi_i\partial\xi_j} = \frac{1}{\Delta_i\Delta_j}\{f(\theta + \Delta_i\xi_i + \Delta_j\xi_j) - f(\theta + \Delta_i\xi_i) - f(\theta + \Delta_j\xi_j) + f(\theta)\} \qquad (18)$$

or, more accurately,

$$\frac{\partial^2 f(\theta)}{\partial\xi_i\partial\xi_j} = \frac{1}{4\Delta_i\Delta_j}\{f(\theta + \Delta_i\xi_i + \Delta_j\xi_j) - f(\theta - \Delta_i\xi_i + \Delta_j\xi_j) - f(\theta + \Delta_i\xi_i - \Delta_j\xi_j$$

$$+ f(\theta - \Delta_i \xi_i - \Delta_j \xi_j)\} \tag{19}$$

The number of times f must be evaluated to obtain the gradients in all directions at one point θ is given in Table 1. The step Δ_i in the difference approach should trade off the truncation error (which increases with Δ_i) and the round-off error (which decreases with Δ_i) (see Stewart [8] for details).

TABLE 1 Number of $f(\theta)$ Function Evaluations Required to Determine Gradients of $J(\theta)$ Using the Finite Difference Approach

NO. OF PARAMETERS m	ONE-SIDED APPROXIMATION FIRST GRADIENT	ONE-SIDED APPROXIMATION FIRST AND SECOND GRADIENTS	TWO-SIDED APPROXIMATION FIRST GRADIENT	TWO-SIDED APPROXIMATION FIRST AND SECOND GRADIENTS
1	2	3	2	3
2	3	6	4	9
3	4	10	6	19
4	5	15	8	33
5	6	21	10	51
10	11	66	20	201
20	21	231	40	801

Computation of ΔM. Computation of the correct Jacobian requires ΔM, which depends on the second derivatives of function f with respect to parameters. This term is small if the outputs have small measurement noise or if the function is mildly nonlinear.

A direct use of Equation (14) to compute ΔM requires the second gradient of f at each iteration. Even the less accurate difference approximation requires many evaluations of $f(\theta)$. These procedures are undesirable from numerical considerations. Nazareth [4] reviews several approaches to determine ΔM without explicit computation of the second gradients of f at each iteration. One technique involves recursively updating second gradients of $f(x_i^T,\theta)$ in each iteration. Note that it is not desirable to update ΔM between iterations, because gradients of $f(x_i^T,\theta)$ behave much better between iterations than does ΔM. The storage requirements are increased, however. For example, with 10 parameters and 500 data points, 10×10 symmetric matrices $\frac{\partial^2 f(x_i,\theta)}{\partial\theta^2}$, $i = 1,2...500$ have to be stored requiring 27,500 words of memory.

Starting Values. Because the nonlinear least-squares problem is iterative, it requires the specification of starting parameter values. Starting values are important because cost functionals may have multiple maxima, saddle points, singular Hessian and discontinuities in the parameter space. Poor behavior may be encountered depending on parameter starting values. From the estimation viewpoint, we are interested in θ which gives the absolute minimum of the fit error. The numerical problems are not characteristic of the least squares approach itself, but are caused by the optimization procedures.

The starting values are specified based on a priori knowledge such that they are close to the final parameter values. If a problem is encountered during the computation, it may be necessary to perturb all the parameters. Sometimes the problem is so severe that a completely new starting value is required. Random search may often precede a gradient optimization algorithm.

III. SYSTEM IDENTIFICATION IN DYNAMIC SYSTEMS

Consider a linear dynamic system, in which the state x (n×1 vector) obeys the differential equation* [8].

$$\dot{x}(t) = (Fx(t) + Gu(t) + \Gamma w(t) \quad 0 \leq t \leq T$$

$$E\{x(0)\} = x_o, \text{ and}$$

$$E\{(x(o) - x_o)(x(o) - x_o)^T\} = P_o \tag{20}$$

and measurements of m linear functions of the state variables are taken at discrete times t_k

$$y(t_k) = Hx(t_k) + v(t_k) \quad k = 1, 2, \ldots, N \tag{21}$$

where u(t) is a q×1 vector of deterministic input and w(t) and $v(t_k)$ are uncorrelated gaussian white noise sources. The power spectral density of w is Q and the covariance matrix of $v(t_k)$ is R. θ is the vector of p unknown parameters in F, G, H, Γ, Q, R, x_o and P_o.

In this paper we only consider the maximum likelihood approach and its various generalizations because they lead to interesting and difficult nonlinear programming problems. Many other methods also require optimizations of criteria similar to the one for maximum likelihood and the techniques presented here will be useful.

In the parameter estimation problems, when the maximum likelihood method is used it is usually more convenient to work with the negative of the logarithm of the likelihood function. It is possible to do so because the logarithm is a monotonic function. It can be shown [9] that the negative log-likelihood function (NLLF) is

*Only linear dynamic models are considered in this paper. Nonlinear dynamic systems require an extended Kalman filter. The nonlinear programming problem, however, is similar.

$$J(\theta) = \frac{1}{2}\sum_{i=1}^{N} \{\nu^T(t_i,\theta)B^{-1}(t_i,\theta)\nu(t_i,\theta)$$

$$+ \log|B(t_i,\theta)|\} \tag{22}$$

where

$$\nu(t_i,\theta) = y(t_i) - E\{y(t_i)|y(t_{i-1}), \ldots$$

$$y(t_2)\ y(t_1)\} \tag{23}$$

and

$$B(t_i,\theta) = E\{\nu(t_i,\theta)\nu^T(t_i,\theta)\} \tag{24}$$

Here, $\nu(t_i,\theta)$ and $B(t_i,\theta)$ denote the innovations and their covariances which may be obtained from the Kalman filter equations for the system ((20) and (21)) (see Appendix A).

The nonlinear programming methods for optimizing negative log-likelihood functions are dictated by the following considerations.

1. NLLF is a sum of two terms: (a) a quadratic function of ν, and (b) a general nonlinear function. Since ν is the difference between a measurement and the expected value of the measurement (Appendix A), the first term resembles the nonlinear least-squares problem.

2. In linear systems in statistical steady state, $B(t_k)$ is not a function of time. Then, the NLLF is first minimized with respect to B to give

$$B = \frac{1}{N}\sum_{i=1}^{N} \{\nu(t_i)\nu^T(t_i)\} \tag{25}$$

and the NLLF for θ becomes

$$J(\theta) = \frac{1}{2}\sum_{i=1}^{N} \left\{ \nu^T(t_i)\hat{B}^{-1}\nu(t_i) + \frac{N}{2}\log|\hat{B}| \right\} \tag{26}$$

Defining

$$\nu'(t_i) = \hat{B}^{-\frac{1}{2}}\nu(t_i) \tag{27}$$

we get

$$J(\theta) = \frac{1}{2}\sum_{i=1}^{N} \nu'(t_i)\nu'^T(t_i)$$

$$+ \frac{N}{2}\log|\hat{B}| \tag{28}$$

The first part is minimized with respect to θ as in nonlinear least-squares.

3. In dynamic systems with continuous measurements or with fast sampling compared to the time constant of the closed-loop Kalman filter, the innovations covariance is nearly equal to measurement noise covariance. The cost functional is then quadratic in innovations.

4. The expected value of the innovations $H\hat{x}(t_i/t_{i-1})$ or its gradients at different time points are not independent of each other. Efficient techniques could be developed to compute the gradients using dynamic formulations. The particular nonlinear programming technique is significantly dictated by the approach used to obtain the gradients of $H\hat{x}(t_i|t_{i-1})$.

5. Selection of parametrization is important in dynamic systems. Kalman gain and innovations covariance lead to more unknown parameters, but the estimation equations are simpler. The process and measurement noise covariance parameters require solutions to fewer, more difficult equations.

6. It is often difficult to specify model form in dynamic identification problems. A reasonable model may be selected prior to maximum likelihood estimation [10]. The model may be further refined during the estimation process. It is therefore necessary to have an estimation procedure which can drop or add parameters (see, e.g., Bierman [3]).

IV. NONLINEAR PROGRAMMING METHODS

Nonlinear programming methods used for minimizing negative log-likelihood functions (NLLF) require gradient computation or NLLF evaluation only. We give a summary of the gradient-based procedures first because many function evaluation methods are implicitly based on gradient methods.

4.1 Gradient-based Nonlinear Programming Methods

A single iteration in gradient-type nonlinear programming methods is

$$M_k(\theta_{k+1} - \theta_k) = \rho_k g_k \tag{29}$$

where θ_k is the parameter vector at the ith iteration, g_k is a vector of gradients of the negative log-likelihood function $J(\theta)$, i.e.,

$$g_k = \left.\frac{\partial J}{\partial \theta}\right|_{\theta = \theta_k} \tag{30}$$

M_k is an approximation to the second partial matrix

$$\left.\left(\frac{\theta^2 J}{\partial \theta^2}\right)\right|_{\theta = \theta_k} \tag{31}$$

and ρ_k is a scalar step-size parameter chosen to ensure that $J(\theta_{k+1}) < J(\theta_k) - \varepsilon$, where ε is a positive number that can be chosen in a variety of ways (see Polak [11]). The class of nonlinear programming methods to be discussed here differs mainly in their selection of M_k, and in some cases ρ_k and g_k. It is shown in Luenberger [12] that the convergence rate near the minimum with ρ_k chosen by a one-dimensional search is

$$J(\theta_{k+1}) \leq \left(\frac{\mu_{max} - \mu_{min}}{\mu_{max} + \mu_{min}}\right)^2 J(\theta_k) \tag{32}$$

where μ_{max} and μ_{min} are the maximum and minimum eigenvalues of $M_k^{-1}(\partial^2 J/\partial\theta^2)$. It is clear from Equation (32) that the best convergence is achieved by making M_k as nearly as possible equal to $(\partial^2 J/\partial\theta^2)$.

Newton-Raphson (NR) Method. In this method M_k is chosen as

$$\left(\frac{\partial^2 J}{\partial\theta^2}\right)\Bigg|_{\theta = \theta_k} \tag{33}$$

and $\rho_k = 1$ except when this choice of ρ_k gives an increase in cost. The convergence is quadratic; however, the method had the following drawbacks: (1) It fails to converge whenever $(\partial^2 J/\partial\theta^2)$ is not positive definite; (2) if $\partial^2 J/\partial\theta^2$ is nearly singular, there are numerical problems in solving (29); and (3) generally, the computation of $\partial^2 J/\partial\theta^2$ is time-consuming (see Section II). Therefore, the NR method is generally not used in parameter estimation problems.

Gauss-Newton (GN) Method. In this method one chooses M_k as the Fisher information:

$$M_k = E\left[\frac{\partial^2 J}{\partial\theta^2}\Bigg|_{\theta = \theta_k}\right] = E\left[\left(\frac{\partial J}{\partial\theta}\right)\left(\frac{\partial J}{\partial\theta}\right)^T\Bigg|_{\theta = \theta_k}\right] \tag{34}$$

The expectation is taken over the whole sample space. M_k is a nonnegative definite symmetric matrix. In statistical literature the above technique is known as the 'Method of Scoring' [13] and in control literature it has been called Modified Newton-Raphson, Quasilinearization and Differential Corrections in somewhat different contexts.

This method is particularly suitable for the singular value decomposition approach of Section I. The upper triangular square root of M_k may be obtained iteratively. Thus the matrix M_k is constrained to be positive definite. Nevertheless, the parameter step in nearly singular directions could be large, violating the locally quadratic cost functional assumption. Levenberg-Marquardt or rank-deficient techniques are then required.

Levenberg-Marquardt Method. In this method M_k is selected as

$$M_k = \left(\frac{\partial^2 J}{\partial\theta^2}\right)\Bigg|_{\theta = \theta_k} + \alpha_k A_k \tag{35}$$

or

$$M_k = E\left[\frac{\partial J}{\partial\theta}\left(\frac{\partial J}{\partial\theta}\right)^T\right]\Bigg|_{\theta = \theta_k} + \alpha_k A_k \tag{36}$$

where A_k is a positive-definite matrix and $\alpha_k > 0$ is a scalar parameter. Generally $A_k = I$ and α_k are chosen large enough so that the eigenvalues of $(M_i + \alpha_i A_i)$ are all positive and above a threshold value. Rules for the selection of α_i are given by Marquardt [14], Bard [15], and Golub etal. [16]. The method has been used successfully in solving nonlinear least-squares problems and is analogous to ridge regression [17].

The Levenberg-Marquardt procedure may be combined with the orthogonal triangularization procedure of Section I by setting

$$U = \sqrt{\alpha_k}\, A_k \tag{37}$$

$$Z = 0$$

before any data point is processed.

Rank-Deficient Solution. The rank-deficient solution is obtained as follows:

$$M_k = \sum_{j=1}^{m} \lambda_j v_j v_j^T \tag{38}$$

$$\Delta\theta_k = \sum_{j=1}^{m} \frac{\rho_k}{\lambda_j}(v_j^T g_k) v_j \tag{39}$$

where λ_j is an eigenvalue of M_k corresponding to the eigenvector v_j. The step size

$$\frac{\rho_i}{\lambda_j}(v_j^T g_i)$$

in direction v_j may be very large for small λ_j. Let the eigenvectors be arranged such that $\lambda_1 > \lambda_2 \ldots \lambda_{m-k} > b > \lambda_{m-k+1} \ldots > \lambda_m$, where b is a suitable threshold. The rank-deficient update is

$$\Delta\theta_k = \rho \sum_{j=1}^{m-k} \frac{v_j^T g_k}{\lambda_j} v_j \tag{40}$$

This will give the estimate in a subspace of θ. In an improved procedure, a gradient search follows rank-deficient estimation.

The method is computationally undesirable because of the need to find eigenvalues and eigenvectors of M_k. If the eigenvectors do not change much from iteration to iteration, M_k can be modified by adding the term

$$\alpha_k \sum_{j=m-k+1}^{m} v_j v_j^T \tag{41}$$

where α_k is some large number. This correction term may be included in the standard procedure.

Scaling Considerations. In system identification problems, the parameters often have different dimensions. This produces an ill-behaved M. Experience has shown that the nondimensional M, resulting from the following transformation, often has better conditioning than the dimensional M.

$$M^*(j,k) = \frac{M(j,k)}{|\theta(j)||\theta(k)|} \tag{42}$$

$$g^*(j) = \frac{g(j)}{|\theta(j)|} \tag{43}$$

where $\theta(j)$ is the nominal value of the jth parameter.

Variable Metric Methods. Since the original work of Davidon [18] and Fletcher and Powell [19], many variable metric methods have been proposed. The main advantage of these methods is that they do not require explicit computation of the Hessian. Instead, these methods update the Hessian or its inverse numerically from gradient information during the search procedure. A comparison of different Variable Metric Methods on nonlinear least-squares problems by Bard [15] shows that the Rank One Correction methods are better than the Davidon-Fletcher-Powell method.* We describe here a Rank One Correction (ROC) method.

Let

$$\Delta\theta_k = \theta_{k+1} - \theta_k \tag{44}$$

$$\eta_k = g_{k+1} - g_k \tag{45}$$

From Taylor series expansion of

$$g_{i+1} = \left.\frac{\partial J}{\partial \theta}\right|_{\theta = \theta_{i+1}}$$

$$\eta_k = H_k \Delta\theta_k \tag{46}$$

where

$$H_k = \left.\frac{\partial^2 J}{\partial \theta^2}\right|_{\theta = \theta_k} \tag{47}$$

$$\Delta\theta_k = H_k^{-1} \eta_k \tag{48}$$

The ROC method consists of modifying an estimate of the Hessian A_k by a rank one matrix B_k such that $A_{k+1} = A_k + B_k$ converges to H_k^{-1} in m steps for a quadratic. If we require, in addition that Equation (44) hold for A_{k+1}, the only solution for B_k turns out to be

$$B_k = \frac{1}{p_k^T \eta_k} p_k p_k^T \tag{49}$$

where

$$p_i = \Delta\theta_k - A_k \eta_k \tag{50}$$

Notice that no one-dimensional search is required and $A_{m+1} = H^{-1}$ for a quadratic cost function $J(\theta)$.

The matrices A_k are not guaranteed to be positive definite. One way to handle this problem is to compute the eigenvalues of A_k and replace the negative ones by their absolute values. This is analogous to the Greenstadt procedure [20] for the Newton-Raphson method.

4.2 Function Evaluation Methods

Starting with the work of Rosenbrock [21], several methods have been developed to minimize a function of several variables without explicit computation of the derivatives of the function. These methods may be divided into two categories:

Explicit Computation of the Gradient. Most of the gradient techniques can be converted into nonderivative methods by using a difference approximation for the gradients of the optimizing function. Parameter perturbations used for differencing may be determined using Stewart's criteria [8] (discussed earlier with reference to $\partial f/\partial\theta$ computation). Once the gradients are specified, the algorithms discussed in Section 4.1 are used. The number of function evaluations for each iteration may be determined from Table 1. These methods essentially emulate gradient-based methods.

*The comparison by BARD [15] of Gauss-Newton and Rank One Correction methods neglects the fact that the gradient alone can be calculated with less computation using adjoint equations (see Section VI).

Direct Search Methods. Rosenbrock [21] suggested the first reasonable direct search algorithm. The algorithm performs a series of one-dimensional searches to find the extremum of a function of several variables. Each coordinate direction is searched by changing one parameter at a time. In the following iteration, the first search direction is replaced by parameter vector correction from current iteration and the remaining search directions are obtained by orthogonalization. The first search direction is efficient but the remaining ones are often inefficient.

In 1964, Powell developed an iterative nonderivative function-minimization technique more efficient than previous approaches [22]. Each iteration of the procedure commences with a search along n linearly independent directions $\xi_1, \xi_2, \ldots, \xi_n$. Initially, these directions are along the coordinate axis. In each iteration, a new direction ξ is defined and the search directions are changed to $\xi_2, \xi_3, \ldots, \xi_n, \xi$. ξ is defined such that the last k of the n directions chosen for (k+1)st iteration are mutually conjugate, if a quadratic is minimized. After n iterations, all directions are mutually conjugate. Sometimes the procedure chooses directions that are linearly dependent. (This is related to overparametrization or the poor conditioning of the Hessian.) If the procedure is continued, minimization will be attained in a subspace of the parameters.

Powell gives a modification of the above procedure in which the search directions are changed to ensure linear independence. A previously chosen direction may be removed under certain circumstances, so that more than n iterations are required to obtain the minimum of a quadratic function (Powell [22]). Nazareth [23] has studied the possibility and consequence of cycling a certain subset of the directions. Zangwill [24] suggests a simplification of the modified Powell procedure. Zangill's procedure converges faster than Powell's basic procedure.

Powell [25] suggested an algorithm to minimize a function which may be represented as the sum of squares of several functions.

V. COMPUTATION OF THE GRADIENT AND HESSIAN OF THE COST FUNCTION

The cost function of Equation (22) depends on $\nu(t_i,\theta)$ and $B(t_i,\theta)$, which are obtained by solving a set of difference-differential equations (Appendix A). There are two different techniques for computing the gradient and Hessian of the negative log-likelihood function: (1) dynamic programming formulation, and (2) sensitivity function propagation.

5.1 Dynamic Programming Formulation [26]

The optimization of the likelihood function may be considered a Kalman filter trajectory control problem with constant control variable θ, since NLLF may be written as

$$V(\hat{x},\theta,t) = \frac{1}{2}\int_t^T [\nu^T(s,\theta)B^{-1} \ldots$$

$$(s,\theta)\nu(s,\theta)$$

$$+ \ln|B(s,\theta)|\delta(s - t_i)]ds \tag{51}$$

where $\delta(s - t_i)$ is the delta function.

Clearly

$$V(\hat{x},\theta,T) = 0$$

$$V(\hat{x}_o,\theta,0) = J(\theta)$$

Define

$$L(\hat{x},\theta,t) = \frac{1}{2}[\nu^T(t,\theta)B^{-1}(t,\theta)\nu(t,\theta)$$

$$+ \ln|B(t,\theta)|\ \delta(t - t_i)] \tag{52}$$

It is easy to show that

$$\frac{\partial V}{\partial t} = -\left(\frac{\partial V}{\partial \hat{x}}\right)^T f - L \tag{53}$$

where

$$f(\hat{x},\theta,t) = \dot{\hat{x}} = (F - KH)\hat{x} + Gu + Ky \tag{54}$$

$$K = K(t)\delta(t - t_i)$$

Equation (53) may now be differentiated with respect to θ to obtain the necessary derivatives. Derivatives of $J(\theta)$ are written as follows:

$$\frac{d}{dt}\frac{\partial V}{\partial \hat{x}} = -\left(\frac{\partial V}{\partial \hat{x}}\right)^T \frac{\partial f}{\partial \hat{x}} - \frac{\partial L}{\partial \hat{x}} \tag{55}$$

$$\frac{d}{dt}\frac{\partial^2 V}{\partial \hat{x}^2} = -\frac{\partial^2 V}{\partial \hat{x}^2}\frac{\partial f}{\partial x} - \left(\frac{\partial f}{\partial \hat{x}}\right)^T \frac{\partial^2 V}{\partial \hat{x}^2} - \frac{\partial^2 L}{\partial \hat{x}^2} \tag{56}$$

$$\frac{d}{dt}\left(\frac{\partial^2 V}{\partial \hat{x}\partial\theta(k)}\right)^T = -\left(\frac{\partial^2 V}{\partial \hat{x}\partial\theta(k)}\right)^T \frac{\partial f}{\partial \hat{x}}$$

$$- \left(\frac{\partial f}{\partial\theta(k)}\right)^T \frac{\partial^2 V}{\partial\hat{x}^2}$$

$$- \left(\frac{\partial V}{\partial \hat{x}}\right)^T \frac{\partial^2 f}{\partial \hat{x}\partial\theta(k)} - \left(\frac{\partial^2 L}{\partial \hat{x}\partial\theta(k)}\right)^T \tag{57}$$

Defining $\lambda = \frac{\partial V}{\partial \hat{x}}$ and $\Lambda_k = \frac{\partial^2 V}{\partial \hat{x}\partial\theta(k)}$,

$$k = 1, 2, \ldots N$$

$$\frac{\partial J}{\partial\theta(j)} = \int_o^T \left[-\lambda^T \frac{\partial f}{\partial\theta(j)} - \frac{\partial L}{\partial\theta(j)}\right] dt \tag{58}$$

$$\frac{\partial^2 J}{\partial\theta(J)\quad(k)} = \int_o^T \left[-2\,{}_k^T \frac{\partial f}{\partial\theta(j)} \right.$$

$$\left. - \lambda^T \frac{\partial^2 f}{\partial\theta(j)\partial\theta(k)} - \frac{\partial^2 L}{\partial\theta(j)\partial\theta(k)} \right] dt \tag{59}$$

5.2 Sensitivity Functions Method

The first and the second gradients of $J(\theta)$ can be computed in terms of the innovations gradients (see Appendix A). This section discussed efficient methods for innovation gradients computations.

Consider first the following system with no process noise.

$$\dot{x} = Fx + Gu \quad x(0) = x_o \tag{60}$$

The system starts from the initial state zero.

The state sensitivity for parameter $\theta(j)$ follows the differential equation

$$\frac{d}{dt}\frac{\partial x}{\partial\theta(j)} = F\frac{\partial x}{\partial\theta(j)} + \frac{\partial F}{\partial\theta(j)}x$$

$$+ \frac{\partial G}{\partial\theta(j)}u \tag{61}$$

$$\frac{\partial x}{\partial\theta(j)}(0) = 0$$

and the innovations sensitivity is

$$\frac{\partial\nu}{\partial\theta(j)} = -H\frac{\partial x}{\partial\theta(j)} - \frac{\partial H}{\partial\theta(j)}x \tag{62}$$

The state sensitivities for all parameters θ can be written as

$$\dot{x}_\theta = F_\theta x_\theta + G_\theta u \tag{63}$$

$$x_\theta(0) = 0$$

where

$$x_\theta = \begin{bmatrix} x \\ \frac{\partial x}{\partial\theta(1)} \\ \cdot \\ \cdot \\ \cdot \\ \cdot \\ \frac{\partial x}{\partial\theta(m)} \end{bmatrix} \quad n(p+1) \times 1 \tag{64}$$

$$F_\theta = \begin{bmatrix} F & & & & \\ \frac{\partial F}{\partial\theta(1)} & F & & 0 & \\ \cdot & 0 & F & & \\ \cdot & \cdot & 0 & \cdot & \\ \cdot & \cdot & \cdot & & \cdot \\ \frac{\partial F}{\partial\theta(m)} & 0 & 0 & & F \end{bmatrix}$$

$$N(p+1) \times n(p+1) \tag{65}$$

$$G_\theta = \begin{bmatrix} G \\ \frac{\partial G}{\partial\theta(1)} \\ \cdot \\ \cdot \\ \cdot \\ \frac{\partial G}{\partial\theta(m)} \end{bmatrix}$$

$$n(p+1) \times q$$

The sensitivity function method requires the solution of $n(p + 1)$ equations (same number of equations is required when there is process noise).

It has been shown that for linear systems, the number of equations required to determine output sensitivity may be substantially decreased [26]. $n(q + 1)$ differential equations must be propagated without process noise and $n(q + m + 1)$ with process noise. Since the number of parameters is often much higher than the number of inputs or measurements, these reductions may give significant savings in computer time.

The first order sensitivities may be used to compute the first gradient as well as an approximation to the second gradient. Therefore, in this approach, an approximation to the second gradient is obtained without any extra differential equations. This unique characteristic of the sensitivity functions makes them extremely useful in nonlinear programming. An exact Hessian will, however, require second gradient of the innovations as well.

VI. COMPARISON OF NONLINEAR PROGRAMMING METHODS

Almost all of the nonlinear programming methods discussed in the previous sections have been used in system identification.

The direct search methods (function evaluation methods) that do not emulate a gradient procedure have been found useful in only simple problems, though the programming requirements make them attractive. The likelihood function often behaves poorly and direct search methods tend to fail. In addition, the computation time requirements are high since the characteristics of the dynamic problem are not used (see, for example, dynamic programming formulation). Therefore, these techniques have not been used extensively. Almost all of the gradient methods can be emulated by a difference approach with reduced programming and computation time requirements (see below).

The number of required differential equations in each of the other methods for computing the first and the second gradients of $J(\theta)$ is given in Table 2. The following conclusions may be drawn from the table:

1. Dynamic programming is the fastest way to get the first derivative of $J(\theta)$.

TABLE 5 Number of Differential Equations for Computing NLLF Gradients

TECHNIQUE	NO. OF DIFFERENTIAL EQUATIONS		REMARKS
	FOR FIRST GRADIENT	FOR FIRST AND SECOND GRADIENTS	
FIRST ORDER SENSITIVITY FUNCTIONS PROPAGATION (GENERAL SYSTEM)	n(p+1)	n(p+1)	APPROXIMATE SECOND GRADIENT
FIRST ORDER SENSITIVITY FUNCTION PROPAGATION (LINEAR SYSTEM)	n(m+q+1)	n(m+q+1)	
SECOND ORDER SENSITIVITY FUNCTION PROPAGATION (GENERAL SYSTEM)	$\frac{n}{2}$ (p+1)(p+2)	$\frac{n}{2}$ (p+1)(p+2)	EXACT SECOND GRADIENT
SECOND ORDER SENSITIVITY FUNCTION PROPAGATION (LINEAR SYSTEM)	n(2m+2q+1)	n(2m+2q+1)	
DYNAMIC PROGRAMMING FORMULATION	2n	$\frac{n}{2}$ (n+2p+5) (DIRECT) 2np (DIFFERENCE APPROX)	
DIRECT DIFFERENCE APPROXIMATION (ONE-SIDED)	n(p+1)	$\frac{n}{2}$ (p^2+3p+2)	

n = STATES, m = MEASUREMENTS, q = INPUTS, p = PARAMETERS.

2. If approximate value of the second gradient is sufficient, the sensitivity function propagation methods are the fastest (for both linear and nonlinear systems).

3. If the measurement noise is high (poor signal-to-noise ratio) exact second gradient is required. For general nonlinear systems, dynamic programming formulation may be used.

4. The number of differential equations for linear systems does not depend on the number of parameters if $p \geq m+q$.

5. In linear systems, even if exact second gradient is desired, sensitivity function propagation is usually the fastest.

It can be seen that the particular method chosen for the computation of cost functional gradients depends on (1) signal-to-noise ratio, (2) linear or nonlinear system, and (3) first or second gradient optimization methods. In each of the above methods, the dynamic nature of the system is used to reduce the number of differential equations propagated.

Experience has shown that difference approximation should be used whenever possible in the propagation of the gradient equations, particularly in system identification with complicated nonlinear differential equations. This reduces programming requirements as well as computation time. Our experience indicates that a new model can be programmed using this method in one-half to as little as one-tenth of the time needed to implement analytical gradient equations. The gradient equations are difficult to check out for programming bugs because of their nonphysical nature.

VII. APPROXIMATIONS

The nonlinear programming methods of the previous section may require unacceptable computation time with long data, many parameters and high-order models. The following approximations provide means for further reductions in computer time.

1. The first gradient may be computed with 2n equations while the computation of even the approximate second gradient requires at least n(m + q + 1) equations. The determination of the second gradient is, in fact, a major part of the computation. Since the only requirement on uniform convergence is that M_k be positive definite, the second gradient matrix could be kept constant between iterations. Our experience shows that the second gradient matrix should be updated at least every fifth iteration with many parameters, more often with fewer parameters. Even though this method requires more iterations, the total computation time is often reduced. This method is particularly suited to the dynamic programming formulation.

2. It is often possible to divide the parameter space into subsets such that parameters in any subset are not correlated with parameters in other subsets. In the first few iterations each set of parameters is optimized separately. When a reasonable convergence is attained, a few iterations with all parameters are performed.

3. Only a short segment of data should be used in the first few iterations. The amount of data is increased as convergence occurs.

4. All the computations should be done less accurately when far from the optimum. For example, an arbitrary stabilizing Kalman filter gain may be selected instead of the optimal filter. The integration step size may be large.

5. An approximate form for the first two gradients may be used in earlier iterations. This reduces the computation time requirements significantly.

VIII. CONCLUSIONS

This paper describes nonlinear programming problems in system identification. The advantages and disadvantages of various techniques used for optimization of the likelihood function are presented. It is hoped that this paper will aid further development and refinement of numerical techniques for dynamic system identification.

IX. ACKNOWLEDGMENTS

The author wishes to thank Dr. Thomas L. Trankle of Systems Control, Inc. for a careful review of the manuscript.

APPENDIX A
INNOVATIONS REPRESENTATION OF DYNAMIC SYSTEMS

The innovations representation is a combination of prediction and measurement update equations.

Prediction

$$\frac{d}{dt}\hat{x}(t/t_{j-1}) = F\hat{x}(t/t_{j-1}) + Gu(t), \quad \hat{x}(t_o|t_o) = x_o$$

$$\frac{d}{dt}P(t/t_{j-1}) = FP(t/t_{j-1}) + P(t/t_{j-1})F^T + \Gamma Q \Gamma^T,$$

$$P(t_o|t_o) = P_o \quad t_{j-1} \le t \le t_j \qquad (A.1)$$

Measurement Update

$$K(t_j) = P(t_j|t_{j-1})H^T(HP(t_j|t_{j-1})H^T + R)^{-1}$$

$$\hat{x}(t_j|t_j) = \hat{x}(t_j|t_{j-1}) + K(t_j)\{y(t_j) - H\hat{x}(t_j|t_{j-1})\}$$

$$P(t_j|t_j) = \{I - K(t_j)H\}P(t_j|t_{j-1}) \qquad (A.2)$$

The innovations and its covariance are related to the above variables

$$\nu(t_j) = y(t_j) - H\hat{x}(t_j|t_{j-1})$$

$$B(t_j) = HP(t_j|t_{j-1})H^T + R \qquad (A.3)$$

The negative log-likelihood function is

$$J(\theta) = \frac{1}{2}\sum_{i=1}^{N}\left\{\nu^T(t_i,\theta)B^{-1}(t_i,\theta)\nu(t_i\theta) + \log|B(t_i,\theta)|\right\}, \qquad (A.4)$$

Its first gradient may be written as

$$\left.\frac{\partial J}{\partial\theta(k)}\right|_{\theta=\theta_i} = \sum_{t=1}^{N}\left\{\nu^T B^{-1}\frac{\partial\nu}{\partial\theta(k)} - \frac{1}{2}\nu^T B^{-1}\frac{\partial B}{\partial\theta(k)}B^{-1}\nu + \frac{1}{2}\mathrm{Tr}(B^{-1}\frac{\partial B}{\partial\theta(k)})\right\}\Bigg|_{\theta=\theta_i} \qquad (A.5)$$

where the arguments of ν and B are not written explicitly and $\theta(j)$ is the jth component of θ vector. In the Gauss-Newton method M_i is generally estimated from the sample as

$$\hat{M}_i(j,k) = \sum_{t=1}^{N}\left\{\left(\frac{\partial\nu}{\partial\theta(j)}\right)^T B\text{-}1\frac{\partial\nu}{\partial\theta(k)} + \frac{1}{2}\mathrm{Tr}\left[B^{-1}\frac{\partial B}{\partial\theta(j)}B^{-1}\frac{\partial B}{\partial\theta(k)}\right] + \frac{1}{4}\mathrm{Tr}\left(B^{-1}\frac{\partial B}{\partial\theta(j)}\right)\mathrm{Tr}\left(B^{-1}\frac{\partial B}{\partial\theta(k)}\right)\right\}\Bigg|_{\theta=\theta_1} \qquad (A.6)$$

An exact expression for M_i derived in Reference [26] may also be used. Moreover, it can be precomputed for a given value of θ. Notice that $\hat{M}$ does not require calculation of $\frac{\partial^2\nu}{\partial\theta(j)\partial\theta(k)}$ and $\frac{\partial^2 B}{\partial\theta(j)\partial\theta(k)}$.

REFERENCES

1. Golub, G. H. (1969). Matrix Decompositions and Statistical Calculations. Statistical Computation edited by R. C. Milton and J. A. Nelder, Academic Press, New York, pp. 365-397.

2. Gill, P. E. and W. Murray (1976). Nonlinear Least Squares and Nonlinearly Constrained Optimization. Lecture Notes in Mathematics No. 506, Springer-Verlag, Berlin.

3. Bierman, G. J. (1977). Factorization Methods for Discrete Sequential Estimation. Mathematics in Science and Engineering, Vol. 128, Academic Press, New York.

4. Nazareth, L. (April 1976). Some Recent Approaches to Solving Large Residual Nonlinear Least Square Problems. Computer Science and Statistics, Ninth Annual Symposium, Cambridge, Mass.

5. Dennis, J. E. (1977). Nonlinear Least Squares and Equations. The State of the Art of Numerical Analysis, edited by D. Jacobs, Academic Press, New York.

6. Golub, G. H., and V. Pereyra (1973). The differentiation of pseudoinverses and nonlinear least square problem whose variables separate. SIAM Journal of Numerical Analysis, Vol. 10, pp. 413-432.

7. Kaminski, P. G., A. E. Bryson, and S. F. Schmidt (Dec. 1971). Discrete Square Root Filtering - A Survey of Current Techniques. IEEE Trans. Auto. Control, Vol. 16, No. 6, p. 727.

8. Stewart, G. W. (1967). A Modification of Davidon's Minimization Method to Accept Difference Approximation of Derivatives. J. ACM 14, pp. 72-83.

9. Hall, W. E., Jr., N. K. Gupta, and R. G. Smith (March 1974). Identification of Aircraft Stability and Control Derivatives for the High Angle-of-Attack Regime. Systems Control, Inc. Report to ONR, 245 pages.

10. Gupta, N. K., W. E. Hall, and T. L. Trankle (May-June 1978). Advanced Methods for Model Structure Determination from Test Data. AIAA Journal of Guidance and Control, Vol. 1, No. 3.

11. Polak, E. (1971). Computational Methods in Optimization, A Unified Approach. Academic Press, New York.

12. Luenberger, D. G. (1972). Introduction to Linear and Nonlinear Programming, Addison Wesley.

13. Rao, C. R. (1965). Linear Statistical Inference and Its Applications. John Wiley and Sons, New York.

14. Marquardt, D. W. (1963). An Algorithm for Least Squares Estimation of Nonlinear Parameters. SIAM J. Num. Anal. 11, pp. 431-441.

15. Bard, Y. (March 1970). Comparison of Gradient Methods for the Solution of Nonlinear Parameter Estimation Problems. SIAM J. Numer. Anal., Vol. 7. No. 1.

16. Golub, G. H., M. Heath, and G. Wahba (Sept. 1977). Generalized Cross-Validation as a Method for Choosing a Good Ridge Parameter. Stanford University Computer Science Department, Report No. STAN-CS-77-622.

17. Marquardt, D. W. (Aug. 1970). Generalized Inverses, Ridge Regression, Biased Linear Estimation, and Nonlinear Estimation. Technometrics, Vol. 12, No. 3, pp. 591-612.

18. Davidon, W. C. (1959). Variable Metric Methods for Minimization. A.E.C. Research and Develop Rep. ANL-5990, Argonne National Lab. Argonne, Illinois.

19. Fletcher, R., and M. J. D. Powell (1963). A Rapidly Convergent Descent Method for Minimization. Comput. J., 6.

20. Greenstadt (1967). On the Relative Efficiencies of Gradient Methods. Math. Comp. 21.

21. Rosenbrock, H. H. (1960). An Automatic Method for Finding the Greatest or Least Value of a Function. Comp. J. 3, p. 175.

22. Powell, M. J. D. (1964). An Efficient Method for Finding the Minimum of a Function of Several Variables Without Calculating Derivatives. Comp. J. 7, pp. 155-162.

23. Nazareth, J. L. (1973). Part I: Unified Approach to Unconstrained Minimization. Part II: Generation of Conjugate Directions for Unconstrained Minimization without Derivatives. Dept. of Computer Science, Univ. of California, Berkeley, Rep. No. 23.

24. Zangwill, W. I. (1967). Minimizing a Function without Calculating Derivatives. Comp. J. 10, pp. 293-296.

25. Powell, M. J. D. (1965). A Method for Minimizing a Sum of Squares of Nonlinear Functions without Calculating Derivatives. The Computer Journal, Vol. 7, p. 303.

26. Gupta, N. K., and R. K. Mehra (Dec. 1974). Computational Aspects of Maximum Likelihood Estimation and Reduction in Sensitivity Functions Computations. IEEE Trans. Auto. Control, Vol. AC-19, pp. 774-783.

COMPARING MATHEMATICAL PROGRAMMING ALGORITHMS BASED ON LAGRANGIAN FUNCTIONS FOR SOLVING OPTIMAL CONTROL PROBLEMS

D. Kraft

Institut für Dynamik der Flugsysteme, DFVLR Oberpfaffenhofen, Federal Republic of Germany

Abstract. A comparison is given of the optimal control application of two highly efficient and conceptually quite different nonlinear programming algorithms based on Lagrangian functions to a rather complex real world aircraft trajectory optimization problem. The algorithms are the Powell-Hestenes-Rockafellar augmented Lagrangian or multiplier method (LMM), and the Biggs-Han-Powell recursive quadratic programming procedure (RQP), respectively. The results are compared to those obtained by applying a generalized reduced gradient algorithm (GRG). It is well known that the efficiency ranking for simple static mathematical programming test problems is (RQP), (GRG), (LMM) (in ascending order). This ranking is no more valid for complicated real life dynamical problems, where especially the gradient evaluations are expensive. In this case (LMM) is marginally faster than (RQP), with (GRG) much less efficient.

Keywords. Optimal control; nonlinear programming; multiplier method; quadratic programming; splines; numerical methods; iterative methods; aerospace trajectories.

1. INTRODUCTION

Since highly efficient mathematical programming algorithms have been developed and implemented to effective and reliable optimization software, within the last ten years, a number of authors have employed them for solving optimal control problems, e.g., Johnson & Kamm (1971), Brusch & Schapelle (1973), Rader & Hull (1975), Sargent & Sullivan (1978), Mantell & Lasdon (1978), Kraft (1978). Among the various nonlinear programming techniques, generalized reduced gradient methods (Lasdon, Sargent), second order quasilinearization methods (Hull), and penalty or multiplier methods in various variants (Johnson, Brusch, Kraft) are used. The purpose of this contribution is to compare nonlinear programming algorithms in case they are applied to solve optimal control problems. Emphasis is on methods based on Lagrangian functions. Especially, to our knowledge, this is the first time that results of the recursive quadratic programming approach used to optimally control dynamic systems are published.

Iterative methods based on Lagrangian functions are known to be very effective for solving constrained optimization problems. The process of refining both their theoretical and numerical properties is most excellently reflected in the survey articles of Fletcher (1974) and Powell (1978a). The development of algorithms is continuing mainly along two conceptually different lines. The first and up to now more familiar technique is the extension of exterior penalty function methods to the method of multipliers (LMM), also known as augmented or penalty Lagrangian method, which has been set forth independently by Hestenes (1969) and Powell (1969) for treating equality constraints, and which has been extended to inequality constraints by Rockafellar (1973). A review of theoretical foundations as well as implementational details can be found in the work of Fletcher (1975); Bertsekas (1976) gives a survey. The other line has rather old roots (Wilson, 1963), but its computational efficiency has only been observed recently by Biggs (1975), Han (1977), and Powell (1978b). The approach is to recursively solve a quadratic program (RQP) with quadratic approximation of the cost function of the problem under consideration and linear approximations of the constraints.

Powell (1978b) has shown the computational superiority of the latter method compared with the former by means of a limited number of standard test examples. The required number of function evaluations is generally reduced by a factor greater than five which compensates by far the expense in solving the quadratic subproblem. Similar trends are given by Schittkowski (1979) in a systematic comparison of a great number of published optimization codes. In this paper results are presented of the computational efficiency of both methods when they are used for the solution of optimal control problems, thereby using a finite dimensional representation of the control functions.

For reference reasons the results are compared to those obtained by solving the problem with a reduced gradient algorithm (Lasdon and colleagues, 1978). In summary it can be stated that much of the advantage of the (RQP) algorithms is lost in this case. The reason is that, though the number of iterations needed to find a minimum of the problem via (RQP) is much less than via (LMM), the computational work to generate the gradients of the involved functions is by far more expensive for (RQP), as will be shown below.

In section 2 of this paper the optimal control problem is stated together with necessary conditions describing an optimal solution. These will be drawn upon in generating gradients of functions. In section 3 a transformation of the original problem to a standard nonlinear programming formulation is proposed, and a method of generating gradients is studied. The multiplier- and recursive quadratic programming methods are recapitulated in section 4. The modelling of the test problem, the dynamics and kinematics of a high performance aircraft, is introduced in the final section which also includes the comparison results.

2. OPTIMAL CONTROL OF A DYNAMIC SYSTEM

2.1 Formulation of the Problem

In a compact interval $[\tau_o,\tau_f]$, τ_o fixed, τ_f generally free, the state vector $x(\cdot): [\tau_o,\tau_f] \to R^n$ of a dynamical system is to be controlled optimally by a control vector $u(\cdot): [\tau_o,\tau_f] \to R^m$, $u\varepsilon C^1$ a.e. $\varepsilon[\tau_o,\tau_f]$, and a parameter vector $v\varepsilon R^p$. It can be assumed that the actual independent variable τ - here the time - is transformed to a normalized time $t = (\tau-\tau_o)/\tau_f$, which ranges in the interval $[0,1]$. In the case of free final time τ_f this can be treated as one element of v.

Consider the optimal control problem

(OCP) minimize the scalar functional

$$\Phi(x,v)_1 \quad , \tag{2.1}$$

subject to the differential constraints

$$\dot{x} - f(x,u,v) = 0 \quad , \tag{2.2}$$

with boundary conditions

$$\begin{aligned} x(0) - x_o &= 0 \quad , \\ \Psi(x,v)_1 &= 0 \quad , \end{aligned} \tag{2.3}$$

and, if necessary, control and state constraints

$$\begin{aligned} C(u,v) &\geq 0 \quad , \\ S(x,v) &\geq 0 \quad , \end{aligned} \tag{2.4}$$

by suitable choice of u and v.

In (OCP) the Mayer formulation of the payoff function with $\Phi(\cdot,\cdot): R^n \times R^p \to R$, $\Phi\varepsilon C^1$, is expressed in (2.1); (2.2) a system of ordinary differential equations, with $f(\cdot,\cdot,\cdot): R^n \times R^m \times R^p \to R^n$, $f\varepsilon C^1$ a.e. in $[0,1]$ for all arguments, describes the dynamic response of the studied system; (2.3) are given initial values x_o of the state vector, and generally nonlinear functional constraints on the final state vector, respectively, with C^1-functions $\Psi(\cdot,\cdot): R^n \times R^p \to R^q$, $0 \leq q \leq n + p$; finally (2.4) with C^1-functions $C(\cdot,\cdot): R^m \times R^p \to R^r$, $0 \leq r \leq m + p$, and $S(\cdot,\cdot): R^n \times R^p \to R^s$, $0 \leq s \leq n + p$, represent control and state constraints, respectively, in the considered interval.

2.2 First Order Necessary Conditions

The calculus of variations supplies conditions necessary for problem (OCP) to be optimal. They are stated here since they will be used in the iterative process later on. Heuristic derivations are given in Bryson & Ho (1969), while proofs from a rather abstract point of view can be found in Neustadt (1976).

To begin with, the Hamiltonian function is defined with the aid of Lagrange multiplier functions $\lambda(\cdot): [0,1] \to R^n$, the adjoint or costate vector,

$$H(x,\lambda,u,v) := \lambda^T f \quad . \tag{2.5}$$

These multipliers have to satisfy the Euler-Lagrange equations

$$\dot{\lambda} + \frac{\partial H}{\partial x} = 0 \quad , \tag{2.6}$$

with boundary conditions given at the end of the interval and arising from the transversality condition

$$\lambda(1) - \left(\frac{\partial \Phi}{\partial x} + \frac{\partial \Psi^T}{\partial x} \nu\right)_1 = 0 \quad , \tag{2.7}$$

$\nu\varepsilon R^q$, a constant multiplier attached to the final condition on the state in (2.3).

Necessary conditions for a control pair (u^*,v^*) to be a stationary point of (2.1) together with the state pair (x^*,λ^*) corresponding to solutions of (2.2) and (2.6) for (u^*,v^*) are:

$$\frac{\partial H}{\partial u} = \frac{\partial}{\partial u} f^T(x^*,u^*,v^*) \cdot \lambda^* = 0 \ ,$$
$$(\frac{\partial H}{\partial v} + \frac{\partial \Phi}{\partial v} + \frac{\partial \Psi^T}{\partial v} \nu)_1 = 0 \ . \qquad (2.8)$$

In problem (OCP) (x,λ) is uniquely defined by (2.2) and (2.6) together with 2n boundary conditions (2.3) and (2.7) with arbitrary (u,v). The multiplier ν is determined by the q additional boundary conditions in (2.3), while m+p conditions for (u^*,v^*) result from (2.8). Thus (OCP) is well determined.

The analysis of necessary conditions has been limited to the case without control and state constraints. The former is easily handled, at least theoretically, by adjoining C with a multiplier functions, say μ to the Hamiltonian, with μ determined from (2.4); the difficulties with the latter are circumvented by defining additional state variables $\dot{x}_{n+i} = \min(S,(x,v),0)$, $(x_{n+i})_o = 0$, $i=1,\dots,s$, while imposing $(x_{n+i})_1 = 0$, thus augmenting the vector Ψ in (2.3) by s elements, a technique which has been used by Kraft (1976) with good success.

3. OPTIMAL CONTROL AS A NONLINEAR PROGRAM

3.1 Transformation of Problem (OCP)

To treat problem (OCP) by nonlinear programming the infinite dimensional control functions u must be represented by a finite set of parameters. While Brusch & Schapelle (1973) and Johnson & Kamm (1971) prefer a polygonial representation and Sargent & Sullivan (1978) use piecewise constant control functions, respectively, Kraft (1978) introduces piecewise polynominal (and piecewise exponential) functions with prescribed continuity conditions.

To be more precise interpolating spline (and exponential spline) functions are taken as basic functions involving a finite number of parameters describing the control in each subinterval $[t_i,t_{i+1})$, $i=1,\dots,\sigma-1$, all of which constitute the interval [0,1]:

$$\Delta = \{0=t_1 < t_2 < \dots < t_\sigma = 1\} \ , \qquad (3.1)$$

the parameter set being

$$\Pi = u_1(t_1),\dots, u_1(t_\sigma), \quad u_2(t_1),\dots, u_m(t_\sigma) \ , \qquad (3.2)$$

resulting in a total number of parameters

$$\nu = m \cdot \sigma + p \ . \qquad (3.3)$$

A cubic spline function S_Δ: $[0,1] \rightarrow R$ on the knot sequence Δ has the following properties

$$S_\Delta \in C^2 [0,1],$$
S_Δ is polynominal of degree three in $[t_i,t_{i+1})$, (3.4)

with the additional requirements

$$S_{\Delta\Pi j}(t_i) = u_j(t_i), \quad i = 1,\dots,\sigma, \ j = 1,\dots,m \ , \qquad (3.5)$$

for interpolating splines $S_{\Delta\Pi}$.

A generalization of (3.1) - (3.5) is furnished by B-splines described in the monography of de Boor (1978), by means of which especially discontinuous controls can easily be treated via multiple knots.

With the set of decision variables

$$y = (u_1(t_1),\dots, u_m(t_\sigma), \quad v_1,\dots, v_p)^T, \ y \in R^\nu \ , \qquad (3.6)$$

the trajectories defined by (2.2) and (2.3) are uniquely defined and can be solved by any efficient numerical method for initial value problems[1]. Thus problem (OCP) can be reformulated in terms of a nonlinear program:
(NLP) minimize the scalar function

$$F(y) \ ,$$
s.t. $\mu = q+r+s$ constraints
$$c_i(y) = 0, \ i \in J = \{i:1,\dots,q+s\} \ ,$$
$$c_i(y) \geq 0, \ i \in K = \{i:q+s+1,\dots,q+r+s\} \ . \qquad (3.7)$$

Of course, (NLP) only approximates (OCP) by the finite representation of u, the approximation being

$$F(y) \cong \Phi(x,v)_1 \ ,$$
$$c_i(y) \cong \Psi_i(x,v)_1, \ i=1,\dots,q,$$
$$c_i(y) \cong S_i(x,v), \ i=q+1,\dots,q+s \ ,$$
$$c_i(y) \cong C_i(u,v), \ i=q+s+1,\dots,q+s+r \ . \qquad (3.8)$$

Note that the differential constraints (2.2) do not enter into formulation (3.7) by solving the initial value problem. An alternative possibility would be to discretize (2.2) and satisfy the resulting equations by collocation.

---

[1] A comparison of inital value problem solvers as used in solving boundary value problems by multiple shooting is given in Dieckhoff and colleagues (1977). In the context of the approach to solve problem (OCP) pursued her the Runge-Kutta-Fehlberg algorithm in the implementation of Shampine & Watts (1976) performed best, with the 4/5 order formulae substituted by 7/8 order formulae.

3.2 Gradients of parametric control models

Both algorithms to be described in the next section will use gradients of functions $g = (F,c_i)^T$, $i=1,\dots,\mu$, involved in (3.7) to solve (NLP). A straightforward way to generate partial derivatives of a function, say g_j, with respect to its arguments y would be by forward differences

$$\frac{\partial g_j}{\partial y_i} = \frac{1}{\Delta y_i}[g_j(y_1,\dots,y_i+\Delta y_i,\dots,y_\nu) - g_j(y_1,\dots,y_i,\dots,y_\nu)] .$$

This is very time-consuming for large ν, because for every parameter disturbancy Δy_i the trajectory (2.2) has to be evaluated. A refined method is to recall the gradient in a function space U, together with consulting the necessary conditions in section 2.2, and to calculate the partials from impulsive response functions, a proposal first given by Brusch & Peltier (1973). Here the result is summarized; a derivation is given in an appendix.

With $\Delta S_{\Delta\Pi} = S_{\Delta\Pi}(y_1,\dots,y_i+\Delta y_i,\dots,y_\nu) - S_{\Delta\Pi}(y_1,\dots,y_i,\dots,y_\nu)$, the disturbance of the parametric control model due to a disturbance Δy_i in any parameter y_i we have:

$$\frac{\partial g_j}{\partial y_i} = \frac{1}{\Delta y_i}\int_0^1 H_u^T \Delta S_{\Delta\Pi} dt, \quad \forall y \in \Pi ,$$
$$\frac{\partial g_j}{\partial y_i} = \int_0^1 H_v \, dt \quad , \forall y \notin \Pi , \qquad (3.9)$$

where H_u and H_v are the deviations from the optimality conditions (2.8) with properly chosen boundary values for the set of λ's in the backward integration of the adjoint system (2.6)

$$\lambda_i^j(1) = \left(\frac{\partial g_j}{\partial x_i}\right)_1, \quad i=1,\dots,n, \ j=0,\dots,\mu,$$

and g_j holding place for the original functions in (3.8).

It should be noted that a routine for generating trajectories should include both possibility (3.9) and forward differences to compute partials, since a comparison of either results indicates the correctnes of coding the adjoint equations (2.6).

4. ALGORITHMS BASED ON LAGRANGIAN FUNCTIONS

4.1 Augmented or penalty Lagrangian

Fletcher (1975) has proposed that to solve problem (NLP) an appropriate penalty function is

$$P(y,\vartheta,S) = F(y) + \frac{1}{2}\gamma(y,\vartheta)^T S\gamma(y,\vartheta),$$
$$\gamma_i(y,\vartheta) = \begin{cases} c_i(y) - \vartheta_i, & \forall i \in J , \\ \min(c_i(y) - \vartheta_i, 0) , & \forall i \in K , \end{cases} \qquad (4.1)$$

$\vartheta \in R^\mu$ and $S = \text{diag}(\sigma_i > 0)$, $\forall i \in J \cup K$. For equality constraints (4.1) reduces to Powell's (1969) function. Elementary evaluation shows the equivalence of (4.1) to Hestenes' (1969) method of multipliers for equality constraints

$$Q(y,\lambda,S) = F(y) - \lambda^T c(y) + \frac{1}{2} c(y)^T Sc(y) ,$$

and to Rockafellar's (1973) generalization to inequality constraints

$$Q(y,\lambda,S) = F(y) - \sum_i \begin{cases} \lambda_i^2/\sigma_i , & \text{if } c_i(y) \ge \lambda_i/\sigma_i \\ (\lambda_i c_i(y) - \frac{1}{2}\sigma_i c_i(y)^2), & \text{else,} \end{cases}$$

both with the relation

$$\lambda = S\vartheta , \qquad (4.2)$$

and the difference between these formulae and (4.1) being $P - Q = 1/2 \cdot \sum_i \lambda_i^2/\sigma_i$, which is independent of y, thus indicating $y(\vartheta,S) = y(\lambda,S)$. The advantage of Fletcher's formulation is the avoidance of the explicit distinction relative to the values of $c_i(y)$.

The aim of the algorithms based on (4.1) is to find minimizing vectors $y(\vartheta,S) \to y^*$, y^* being a local minimizer of (NLP), without forcing $S \to \infty$ for ensuring convergence. The latter is necessary for the simple exterior penalty function ($\vartheta = 0$ in(4.1)) as is well-known. The aim is reached by introducing a master iteration $\vartheta \to \vartheta^*$ with associated λ^* according to (4.2), the vector of Lagrange multipliers at the solution y^* of (NLP), thereby trying to keep S constant at moderate values, and to increase S only in case the rate of convergence of $c(y(\vartheta,S))$ to zero is not sufficiently rapid. The overall algorithm (LMM) consists of the following steps

(i) select y^o, ϑ^o, S^o .

(ii) $y^k(\vartheta,S) = \arg\min_y P(y,\vartheta^k,S^k)$,

(iii) if $\|c^k\|_\infty < \rho\|c^{k-1}\|_\infty$, $0<\rho<1$, then

$$\vartheta_i^{k+1} = \vartheta_i^k - \begin{cases} c_i(y^k) & \forall i \in J, \\ \min(c_i(y^k), \vartheta_i^k), & \forall i \in K, \end{cases}$$
$$S^{k+1} = S^k ,$$

return to (ii); else

(iv) $\vartheta^{k+1} = \vartheta^k$,

$S^{k+1} = \varkappa S^k$, $\varkappa > 1$,

return to (ii) .

Remark 1. It can be shown, see e.g. Fletcher (1975), that the master iteration (step (iii) in algorithm (LMM)) is a steepest ascent step in maximizing the function dual to (4.1) with respect to ϑ

$$\vartheta^{k+1} = \arg\max_{\vartheta} \{\min_{y} P(y,\vartheta)\} ,$$

where S is kept constant and dropped from the argument list. An alternative is a Newton step in maximizing $\min_{\vartheta} P(y,\vartheta)$

$$\text{(iiia)} \quad \vartheta_i^{k+1} = \vartheta_i^k - \begin{cases} E_1, & \forall i\varepsilon J \\ E_2, & \forall i\varepsilon K \wedge c_i(y^k) < \vartheta_i^k \\ \vartheta_i^k, & \forall i\varepsilon K \wedge c_i(y^k) \geq \vartheta_i^k \end{cases} ,$$

where

$$E_1 = (A^T G^{-1} A)^{-1} c_i(y^k)/\sigma_i ,$$

$$E_2 = \min((A^T G^{-1})^{-1} c_i(y^k)/\sigma_i, \vartheta_i^k) .$$

Here A and G denote the Jacobian matrix $[\nabla_x c_1, \ldots, \nabla_x c_\mu]$ of the active constraints and the Hessian matrix $\nabla_x^2 P$ of the penalty Lagrangian, respectively.

Remark 2. Step (ii), the inner iteration, is an unconstrained minimization, and can be solved efficiently by quasi-Newton methods, thereby obtaining an approximation to G^{-1}, which can be used in (iiia). An up-to-date review of quasi-Newton methods using matrix factorizations is given by Gill & Murray (1978).

4.2 Recursive quadratic programming

Han (1977) has proposed the following globally convergent extension of the damped Newton method for solving problem (NLP): generate a sequence $\{y^k\}$ converging to a local solution y^* of (NLP) by means of recursively solving the quadratic programming problem

$$\min_{p} \ \nabla F(y)^T p + \tfrac{1}{2} p^T B p$$

$$\text{s.t.} \quad \begin{aligned} \nabla c_i(y)^T p + c_i(y) &= 0, \ \forall i\varepsilon J, \\ \nabla c_i(y)^T p + c_i(y) &\geq 0, \ \forall i\varepsilon K, \end{aligned} \qquad (4.3)$$

which yields the search direction p in which the next approximation to the solution is found:

$$y^{k+1} = y^k + \alpha_k p^k . \qquad (4.4)$$

Here α_k is a relaxation factor to ensure global convergence; it is gained from a steplength algorithm by reducing properly the exact penalty function

$$R(y,\rho) = F(y) + \sum_{i\varepsilon J} \rho_i |c_i(y)| + \sum_{i\varepsilon K} \rho_i |\min(c_i(y),0)| , \qquad (4.5)$$

with suitably chosen parameters ρ. In (4.3) the $\nu \times \nu$ matrix B is undefined yet, and it is at this point that the relation of the method to Lagrangian functions is introduced, in that B is taken as an approximation to the Hessian matrix of the Lagrangian of (NLP)

$$L(y,\lambda) := F(y) - \lambda^T c(y) ,$$

instead that of F(y) alone. Summarized, algorithm (RQP) includes the following steps

(i) start with y^o, B^o ,

(ii) solve (4.3) for a Kuhn-Tucker pair (p^k, λ^k) ,

(iii) find α_k and y^{k+1} by requiring

$$R(y^k + \alpha_k p^k, \rho^k) < R(y^k, \rho^k) ,$$

(iv) update B^{k+1} by some scheme; return to (ii).

Remark 3. Solving quadratic programs is now standard in the literature, e.g. Fletcher (1971), Gill & Murray (1978). Bunch & Kaufman (1977) have proposed an efficient method for indefinite quadratic programming.

Remark 4. Powell (1978b) has given computationally efficient details for algorithm (RQP):

(1) $\rho_i^k = \max(|\lambda_i^k|, \tfrac{1}{2}(\rho_i^{k-1} + |\lambda_i^k|))$, $i\varepsilon J \cup K$

(2) a modification of the B-F-G-S-update for B^{k+1} to remain positive definite in the constrained case.

5. TEST PROBLEM

5.1 Model of a High Performance Aircraft

To give an example for problem (OCP) and its numerical solution, the 3-dimensional motion of the center of gravity of a non-yawing aircraft above flat, non-rotating earth is considered. The state variables are defined by the following set of differential equations given in an axis system coupled with the flight path (cp. Miele, 1962; Etkin, 1972; Brüning & Hafer, 1978):

$$\dot{V} = \frac{1}{m}(T\cos\alpha - D) - g\sin\gamma ,$$
$$\dot{\chi} = \frac{1}{mV}(T\sin\alpha + L)\frac{\sin\mu}{\cos\gamma} ,$$
$$\dot{\gamma} = \frac{1}{mV}(T\sin\alpha + L)\cos\mu - \frac{g}{V}\cos\gamma ,$$
$$\dot{m} = -C_m T \tag{5.1}$$
$$\dot{\xi} = V\cos\gamma \cos\chi,$$
$$\dot{\eta} = V\cos\gamma \sin\chi,$$
$$\dot{\zeta} = V\sin\gamma .$$

The components of the state vector $x = (V,\chi,\gamma,m,\xi,\eta,\zeta)^T$ are velocity V, trajectory yaw angle χ, trajectory pitch angle γ, mass m, and coordinates ξ,η,ζ of c-g relative to a geodetic axes system, respectively. The aircraft is controlled by control vector $u = (\alpha,\mu)^T$, with components angle of attack α, and trajectory bank angle μ. The aerodynamic forces lift L and drag D depend on the dynamic pressure $q = \frac{1}{2}\rho(\zeta)V^2$ and the aerodynamic parameters C_L and C_D, respectively:

$$L = q\ S\ C_L(M,\alpha) ,$$
$$D = q\ S\ C_D(M,\alpha) , \tag{5.2}$$

where $\rho(\zeta)$ is the atmospheric density and S a reference area. C_L and C_D are generally functions of the angle of attack and Mach number $M = V/a(\zeta)$, with a (ζ) the velocity of sound. Note that both ρ and a change with altitude ζ. Thrust force T and specific fuel consumption C_m are given by

$$T = C_T(M,\zeta)\ T_o ,$$
$$C_m = \bar{C}_m(M,\zeta)\ C_{mo} , \tag{5.3}$$

where index o indicates reference values. Thrust coefficients C_T and $\bar{C}_m$ are functions of Mach number and altitude as well as engine throttle position δ, generally; the latter dependence suppressed here, because only time optimal flights are considered, in which $\delta = 1$ is the optimal throttle position. It should be noticed that for other cost criteria, δ is an additional element of the control vector function u. The functional dependence of the coefficients in (5.2) and (5.3) is quite complicated and is described in more detail in Kraft (1978). In (5.1) g is the acceleration of gravity.

We want to find minimum time trajectories with

$$\Phi = t_1$$

for given initial state conditions

$$x(0) = x_o ,$$

some prescribed final conditions (5.4)

$$x(1) = x_1 ,$$

and linear constraints to the control

$$u_{min} \le u(t) \le u_{max} .$$

Note that the final conditions are by no means linear with respect to u, and y, respectively, by the nonlinearity of the systems equations (5.1). It should be reasonable from (5.1) that the adjoint equations supplementing (5.1) to the Euler-Lagrange equations, and defined by (2.6), are rather lengthy, why we omit them here; instead we rever again to Kraft (1978).

5.2 Test Results and Discussion

The original motivation for problem (5.4) was to investigate whether flight in the post-stall flight region ($\alpha > \alpha_{CLmax}$) might result in a reduction of the cost function Φ. For this reason problem (5.4) has been solved for the boundary values given in table 1 with $\alpha_{max} = 20^o$ (conventional flight) and $\alpha_{max} = 90^o$ (post-stall flight), respectively. The results are summarized in fig. 1 and they indicate a pronounced post-stall time advantage for boundary velocities below 125 m/s. Remarkable is also the fact that below boundary velocities of 100 m/s the required boundary values can only be achieved by post-stall flight. Above boundary velocities of 185 m/s a constraint on the load factor $n = L/(m \cdot g) \le n_{max}$ was activated, which has been treated by the technique described in section 2.2.

The preceding results have all been obtained by applying algorithm (LMM) in our implementation in the code CFMIN, which is an improved version of FUNMIN (Kraft, 1978) mainly by the introduction of Fletcher's (1975) master iteration. These results should serve as (i) reference to the capabilities of nonlinear programming in analyzing a rather complex flight mechanics task and as (ii) introduction to the main item of the contribution, the comparison of the computational efficiency of algorithm (LMM) with respect to algorithm (RQP) in solv-

Table 1 Boundary Values of State Variables in Problem (5.4)

x	V[m/s]	$\chi[^o]$	$\gamma[^o]$	m[kg]	ξ[m]	η[m]	ζ[m]
t_0	from 50 to 200	0	0	18000	1000	1000	2000
t_1	from 50 to 200	-180	0	free	1000	1000	2000

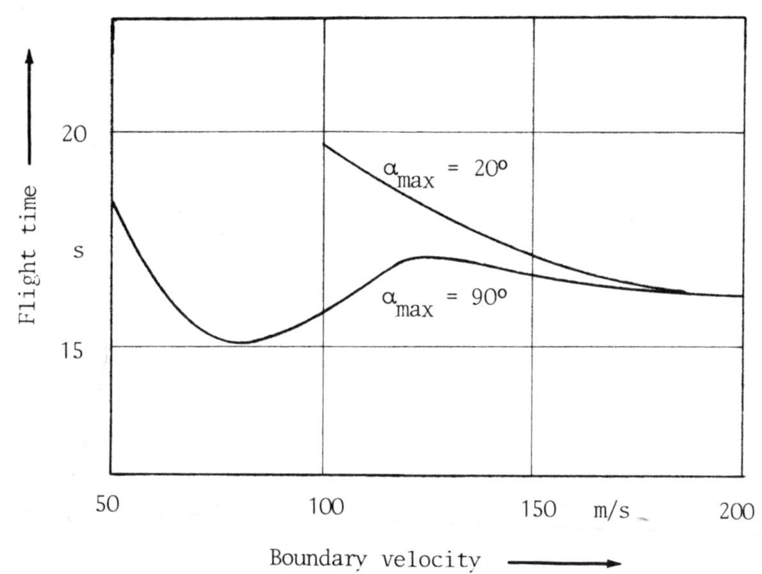

Fig. 1. Flight time τ_f versus boundary velocity $V_o=V_f$.

Table 2 Summary of Results for Algorithm Comparison

No	Algorithm	α_{max}	ICYC	ITER	IFCT	F	$\|\|c\|\|_\infty$	CPU+
1	(LMM)	90	7	190	200	17.5032	10^{-5}	63
2	(RQP)	90		50	86	17.5038	10^{-4}	83
3	(RQP)*	90		50	86	17.5038	10^{-4}	125
4	(LMM)	90	7	256	280	17.5028	10^{-6}	87
5	(LMM)^N	90	7	289	304	17.5028	10^{-8}	97
6	(GRG)	90		99	967/99	17.5026	10^{-5}	161
7	(GRG)	30		119	1155/119	18.9890	10^{-5}	169
8	(LMM)	30	7	123	128	18.9884	10^{-5}	34
9	(RQP)	30		28	30	19.0112	10^{-5}	38

* Original steplength algorithm

N Newton-correction in (LMM) master iteration

\+ CPU in seconds on Amdahl 470 V/6 with Fortran H-extended compiler in optimizing (2) version

ing optimal control problems like problem (5.4). In this case the boundary velocity is restricted to 110 m/s with all other boundary values as in table 1. Instead of $\alpha_{max} = 20^o$ for the conventional flight, $\alpha_{max} = 30^o$ is admitted. The implementation of algorithm (RQP) is Powell's (1978) VFO2AD from Harwell library.

The main results are shown in table 2, where ICYC means the number of master iteration cycles in algorithm (LMM), ITER the total number of iterations, and IFCT the number of function and gradient evaluations, respectively. The tolerances of the algorithms have been chosen such that the resulting accuracy in satisfying the constraints has been approximately the same. Let rest, the concentration for the moment on the case $\alpha_{max} = 90^o$ (lines 1-6 of table 2): 190 iterations of algorithm (LMM) are facing 50 iterations of algorithm (RQP), only slightly more than a quarter! But a comparison of computer time needed nevertheless indicates an advantage for algorithm (LMM): 63 vs. 83. The reason is easily understood if one compares the number of explicit functions involved in both algorithms (see equations (4.1) and (4.3), respectively) for which gradients have to be evaluated: only one in (LMM), but $1 + \mu$ in (RQP), and for each function one set of multiplier functions λ (2.6) has to be generated for the gradient (3.9). This implies that the work to be done in algorithm (RQP) depends directly on the number of boundary values, whereas it is independent of this number for algorithm (LMM). An exception to this is the case when the Newton step (iiia) is used in the master iteration for updating the Lagrange multipliers in algorithm (LMM). Then the partials of all functions involved have to be formed, and the computational expense is the same for both algorithms. But the time for this update is almost neglegible compared to the total time needed, as the number of iterations exceeds by far the number of cycles. It should be noted that the original version of VF02AD would have needed 125 seconds for the same problem, as it did not take into account the possibility to skip gradient evaluations for (IFCT-ITER) function evaluations, the extra function evaluations to find the convergence-ensuring damped Newton steplength. This saving is not possible in algorithm (LMM) because the Armijo-steplength algorithm implemented needs gradient information as well as function information, see Ortega & Rheinboldt (1970). The relationship between the number of iterations and the number of functions indicates that the steplength choice is more critical for (RQP) than it is for (LMM).

Another interesting result for algorithm (LMM) is the influence of the Newton step against the steepest ascent correction in updating the multipliers (cf. lines 4 and 5 of table 2). This case has been chosen to test the accuracy of algorithm (LMM), therefore the tolerance has been raised. The result: the Newton correction causes a better satisfaction of the constraints (two decimals) with moderately increased number of iterations.

Our experience in using the generalized reduced gradient algorithm (GRG) has not been very encouraging in the context of the optimal control problem (cf. line 6 in table 2). The number of overall iterations is about in between (LMM) and (RQP). For each iteration gradients of every function involved in problem (5.4) have to be evaluated together with rather many function evaluations for an apparently optimal steplength choice, resulting in a CPU time of 161 seconds, almost threee times as much as (LMM) needs. One of the reasons for the inefficiency is the expensive restauration phase to start the minimization with a feasible decision vector. This phase took 55 of 99 iterations.

Supplementary results are summarized in lines 7 to 9 of table 2: the case $\alpha_{max} = 30^o$. The same starting vector (α^o,μ^o) has been used as for $\alpha_{max} = 90^o$ for (LMM) and (RQP); and it is evident from the number of iterations needed to solve the problem that this initial vector is "closer" to the 30°-solution than it is to the 90°-solution. For (GRG) the starting vector (α^o,μ^o) had to be modified because with the original (α^o,μ^o) no feasible point could be found by the algorithm.

Some graphical information is presented in figures 2 to 9: the graphs of the optimal controls and resulting state variables, respectively, for both α limitations. Figure 8, the projection of the flightpath onto a horizontal plane shows that the time optimal post stall flight offers also pathlength advantages: the flightpath almost lies in the vertical plane. Figure 9 gives a comparison of the cubic spline parametric control model versus an exponential spline with certain stiffness parameters. The latter model is a better representation at the constraints $\alpha \leq \alpha_{max}$ but the disatvantage is the greater amount of time needed to calculate exponential spline coefficients and to evaluate the spline at the interpolation points.

CONCLUSIONS

A method is presented to transform an optimal control problem to a nonlinear programming problem. Algorithms based on Lagrangian functions to solve the latter are briefly described. It is shown that these methods compare favorably with other well-known optimization routines, based on generalized reduced gradient ideas, for instance. Although this comparison is due to the limited number of test runs by far not representative for the entire spectrum of solving optimal control problems by nonlinear programming methods the results are indicative for the selection of algorithms and software for future application.

Algorithms based on Lagrangian functions considered here are the Lagrange multiplier methods (LMM) and the recursive quadratic programming technique (RQP), showing marginal advantages for the former in the optimal control case. But while algorithmic implementation of the multiplier method is already rather sophisticated some areas for future research seem to be left for recursive quadratic programming, e.g. update of the Hessian matrix of the Lagrangian function, the use of indefinite quadratic programming, or the modification of the cost function in the steplenth algorithm.

ACKNOWLEDGEMENTS

This work constitutes a portion of a larger collaboration in the field of optimal control of aircraft trajectories at the Instiute of Dynamics of Flight Systems of DFVLR. The author is indebted to Prof. R. Bulirsch of TU Munich and to my colleagues K.H. Well, G.C. Shau, and E.G. Berger for many fruitful discussions, valuable suggenstions and help.

A.E.R.E. in Harwell is kindly acknowledged for providing subroutine VF02AD to me, which only made this comparison possible.

REFERENCES

Bertsekas, D. P. (1976). Multiplier methods: a survey. *Automatica 12*, 133-145.

Biggs, M. C. (1975). Constrained minimization using recursive quadratic programming. In L.C.W. Dixon, and G.P. Szegö (Eds.), *Towards global optimization*. North-Holland, Amsterdam.

de Boor, C. (1978). A practical guide to splines. Springer, New York.

Brüning, G. & X. Hafer (1978). Flugleistungen. Springer, Berlin.

Brusch, R. G. & J. P. Peltier (1973). Parametric control models and gradient generation by impulsive response. Proc. IAF-Congress, Baku.

Brusch, R. G. & R. H. Schapelle (1973). Solution of highly constrained optimal control problems using nonlinear programming. *AIAA Journal 11*, 135-136.

Bunch, J. R. & L. Kaufman (1977). Indefinite quadratic programming. Comp. Sci. Techn. Rept. 61, Bell Laboratories.

Bryson, A. E. & Y. C. Ho (1969). Applied optimal control. Ginn & Company, Waltham, Mass..

Dieckhoff, H. J., P. Lory, H. J. Oberle, H. J. Pesch, P. Rentrop and R. Seydel (1977). Comparing routines for the numerical solution of initial value problems of ordinary differential equations in multiple shooting. *Numer. Math. 27*, 449-469.

Etkin, B. (1972). Dynamics of atmospheric flight. John Wiley & Sons, Inc., New York.

Fletcher, R. (1971). A general quadratic programming algorithm. *J. Inst. Maths Applics 7*, 76-91.

Fletcher, R. (1974). Methods related to Lagrangian functions. In P. E. Gill & W. Murray (Eds.), *Numerical methods for constrained optimization*. Academic Press, London. pp. 219-239.

Fletcher, R. (1975). An ideal penalty function for constrained optimization. *J. Inst. Maths Applics 15*, 319-342.

Gill, P. E. & W. Murray (1978). Numerically stable methods for quadratic programming. *Math. Program. 14*, 349-372.

Gill, P. E. & W. Murray (1978). Tutorial on unconstrained minimization. In H. J. Greenberg (Ed.), *Design and implementation of optimization software*. Sijthoff & Nordhoff, Leiden.

Han, S. P. (1977). A globally convergent method for nonlinear programming. *J. Optim. Theory Appl. 22*, 297-309.

Hestenes, M. R. (1969). Multiplier and gradient methods. *J. Optim. Theory Appl. 4*, 303-320.

Johnson, I. L. & J. L. Kamm (1971). Parameter optimization and the space shuttle. Proc. JACC, St. Louis, Mo., 776-781.

Kraft, D. (1976). Optimierung von Flugbahnen mit Zustandsbeschränkungen durch Mathematische Programmierung. DGLR Jahrbuch, München, 201.1-201.23.

Kraft, D. (1978). Nichtlineare Programmierung - Grundlagen, Verfahren, Beispiele. DLR-FB 77-68, DFVLR, Köln.

Lasdon, L. S., A. D. Waren, A. Jain, and M. Ratner (1978). Design and testing of a generalized reduced gradient code for nonlinear programming. *ACM Trans. Math. Softw. 4*, 34-50.

Luenberger, D. J. (1969). Optimization by vector space methods. John Wiley & Sons, Inc., New York.

Mantell, J. B. & L. S. Lasdon (1978). A GRG algorithm for econometric control problems. *Ann. Econ. Soc. Meas. 6*, 581-597.

Miele, A. (1962). Flight mechanics - theory of flight paths. Addison - Wesley, Reading, Mass..

Neustadt, L. W. (1976). Optimization. Princeton University Press, Princeton.

Ortega, J. M. & W. C. Rheinboldt (1970). Iterative solution of nonlinear equations in several variables. Academic Press, New York.

Powell, M. J. D. (1969). A methods for nonlinear constraints in minimization problems. In R. Fletcher (Ed.), *Optimization*. Academic Press, London. pp. 283-298.

Powell, M. J. D. (1978a). Algorithms for nonlinear constraints that use Lagrangian functions. *Math. Program. 14*, 224-248.

Powell, M. J. D. (1978b). A fast algorithm for nonlinearly constrained optimization calculations. In G. A. Watson (Ed.), *Numerical Analysis*. Springer, Berlin. pp. 144-157.

Rader, J. E. & D. G. Hull (1975). Computation of optimal aircraft trajectories using parameter optimization methods. *J. Aircraft 12*, 864-866.

Rockafellar, R. T. (1973). A dual approach to solving nonlinear programming problems by unconstrained optimization. *Math. Program. 5*, 354-373.

Sargent, R. W. H. & G. R. Sullivan (1978). The development of an efficient optimal control package. In J. Stoer (Ed.), *Optimization techniques, Part 2*. Springer, Berlin. pp. 158-168.

Schittkowski, K. (1979). A numerical comparison of 13 nonlinear programming codes with randomly generated test problems. In L. C. W. Dixon & G. P. Szegö (Eds.). *Numerical optimization of dynamical systems*. North-Holland, Amsterdam.

Shampine, L. F. & H. A. Watts (1976). Practical solution of ordinary differential equations by Runge-Kutta methods. Sandia Laboratories Report SAND 76-0585, Albuquerque, New Mexico.

Wilson, R. B. (1963). A simplicial method for convex programming. Ph. D. thesis, Harvard University, Cambridge, Mass..

APPENDIX

In this appendix a derivation of result (3.9) is given, the gradient generation by impulsive response functions for parametric control models.

Let u be an element of the Hilbert Space $U = L_2[a,b]$ with inner product $(v,w) = \int_a^b v(t)^T \cdot w(t)dt$ and let f be a functional defined on U with range in R^1. Then the differential

$$\delta f(u;h) = (f_u,h)$$

is a Fréchet differential, provided that f_u the gradient of f at u exists, Luenberger (1969). Note that the space of all piecewise third order polynominals $S_{\Delta\Pi}$ (equ. (3.5)) is a subspace of U. We want to satisfy the optimally conditions (2.8). Thus deviations, e.g. $H_u \neq 0$, must be forced to zero by changing u. Natural candidates for the differentials of the functionals involved in problem (OCP) are therefore, e.g.,

$$\delta\Phi = \int_0^1 H_u^T \, \delta u \, dt \quad , \tag{A.1}$$

with H defined as in (2.5) and the relation to the considered functionals given by (2.7). Now, define the variation of the control function δu in terms of the variation δy of the decision variables $y \in \Pi$

$$\delta u(t) \cong \delta S_{\Delta\Pi}(y,\delta y) = S_{\Delta\Pi}(y+\delta y) - S_{\Delta\Pi}(y),$$

and replace Φ in (A.1) by F as in (3.8), to get

$$\delta F = \int_0^1 H_u^T \, (S_{\Delta\Pi}(y+\delta y) - S_{\Delta\Pi}(y)) \, dt \quad . \tag{A.2}$$

By taking partials in (A.2) the desired result (3.9-1)

$$\frac{\partial F}{\partial y_i} = \frac{1}{\Delta y_i} \int_0^1 H_u^T \, (S_{\Delta\Pi}(y+\Delta y) - S_{\Delta\Pi}(y)) \, dt \tag{A.3}$$

with $\Delta y = (y_1,\dots,y_i+\Delta y_i,\dots,y_\nu)^T$

is achieved. Note that in (A.3) it is not necessary to consider $\Delta y_i \to 0$ by the additivity property of $S_{\Delta\Pi}$. Similar considerations lead to result (3.9-2) for $y \notin \Pi$, with the difference that δu must be subseded by δv which does not enter into $S_{\Delta\Pi}$.

Remark: There is a small but crucial difference of the results of this appendix compared to those derived by Brusch & Peltier (1973), where for negative δy the partials will be wrong (equ. (4.10), p. 23).

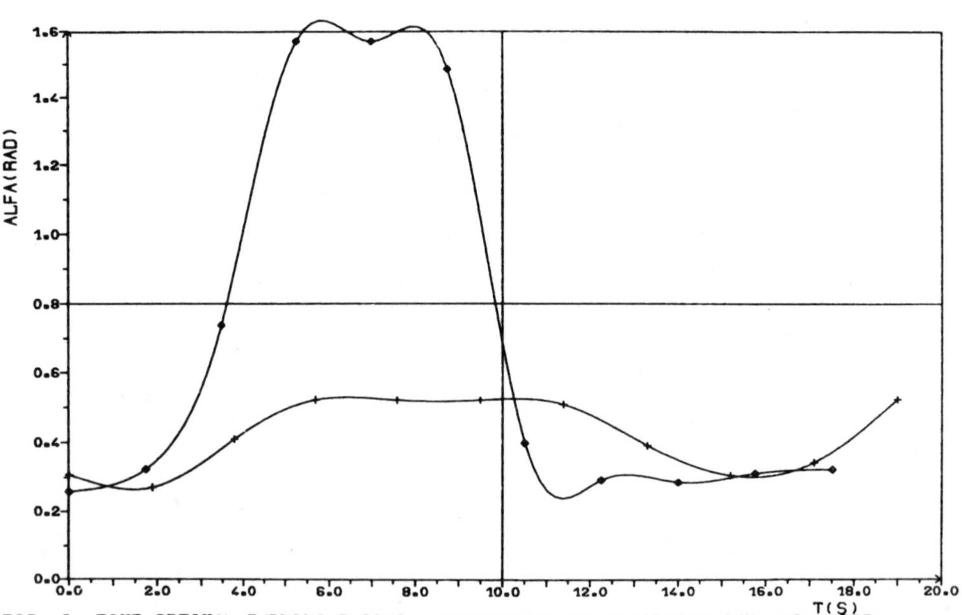

FIG. 2. TIME OPTIMAL TURNING FLIGHT - COMPARISON OF ALGORITHM (MM) VS. (RQP)
GRAPH OF ANGLE OF ATTACK
+ ALFA MAX = 30 ◇ - ALFA MAX = 90

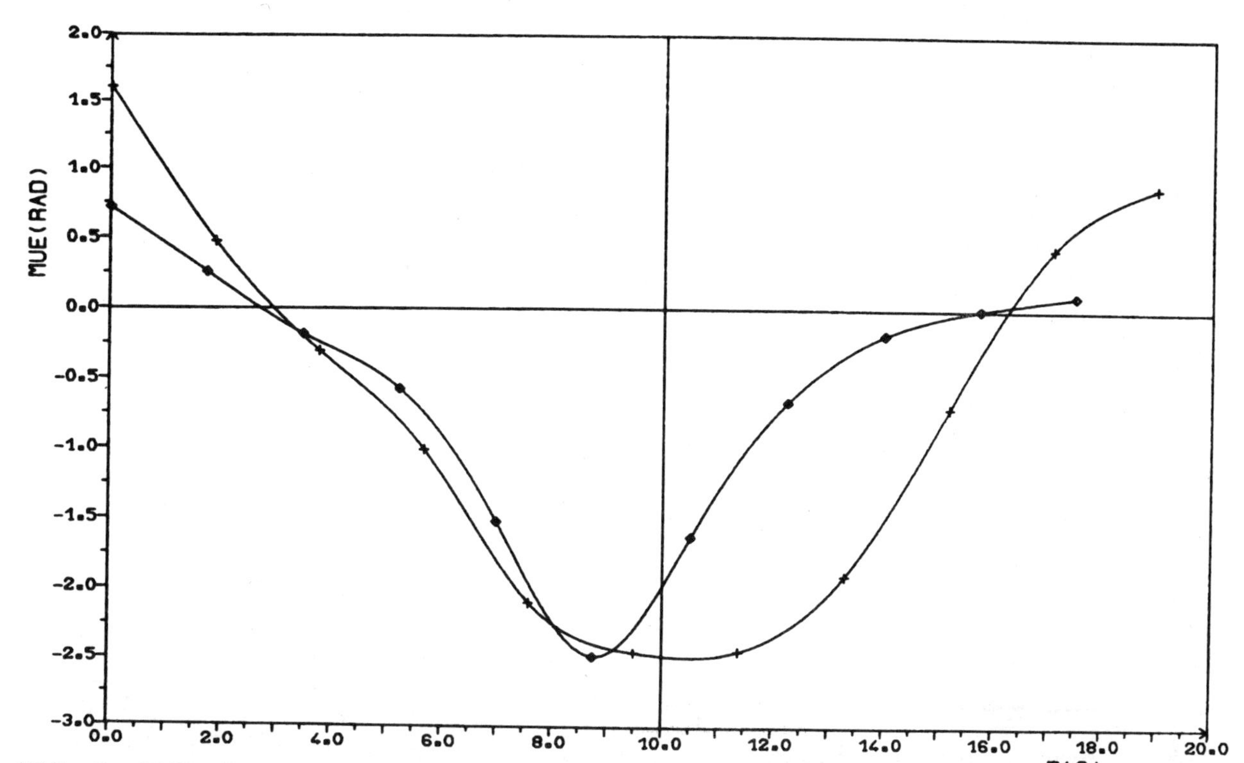

FIG. 3. TIME OPTIMAL TURNING FLIGHT - COMPARISON OF ALGORITHM (MM) VS. (RQP)
GRAPH OF FLIGHTPATH ROLL ANGLE
+- ALFA MAX = 30 ◇- ALFA MAX = 90

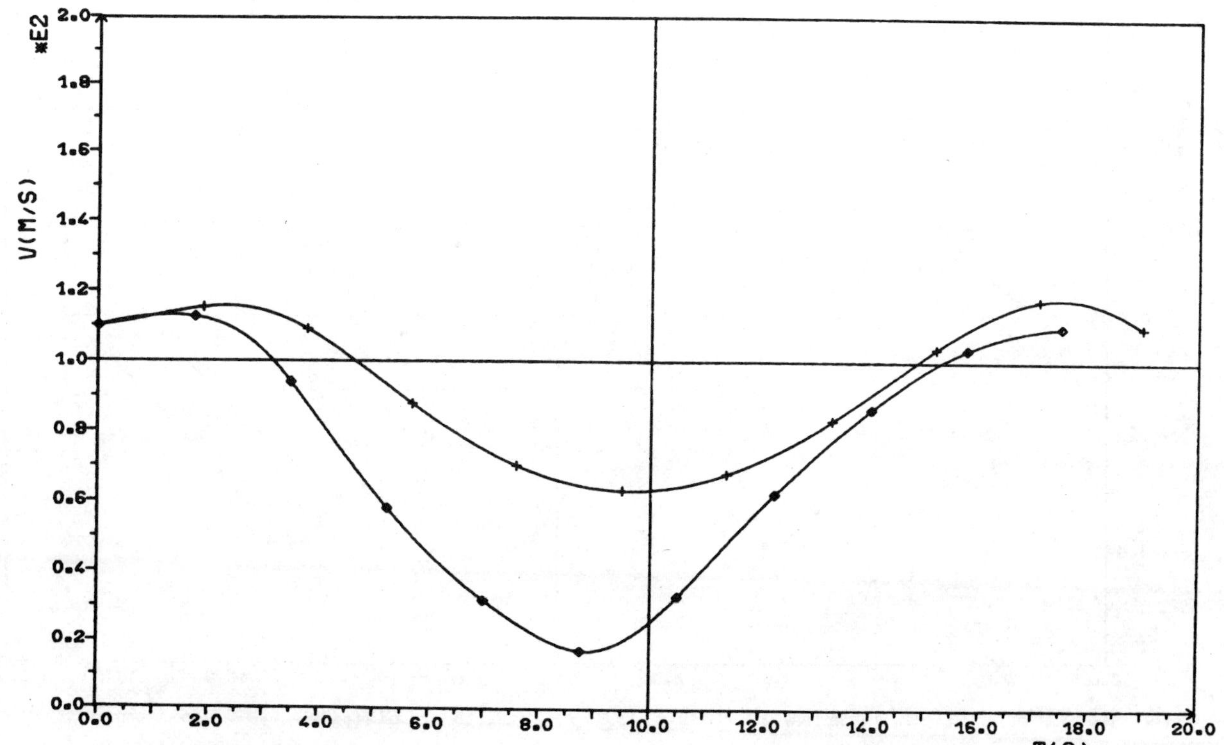

FIG. 4. TIME OPTIMAL TURNING FLIGHT - COMPARISON OF ALGORITHM (MM) VS. (RQP)
GRAPH OF FLIGHT VELOCITY
+- ALFA MAX = 30 ◇- ALFA MAX = 90

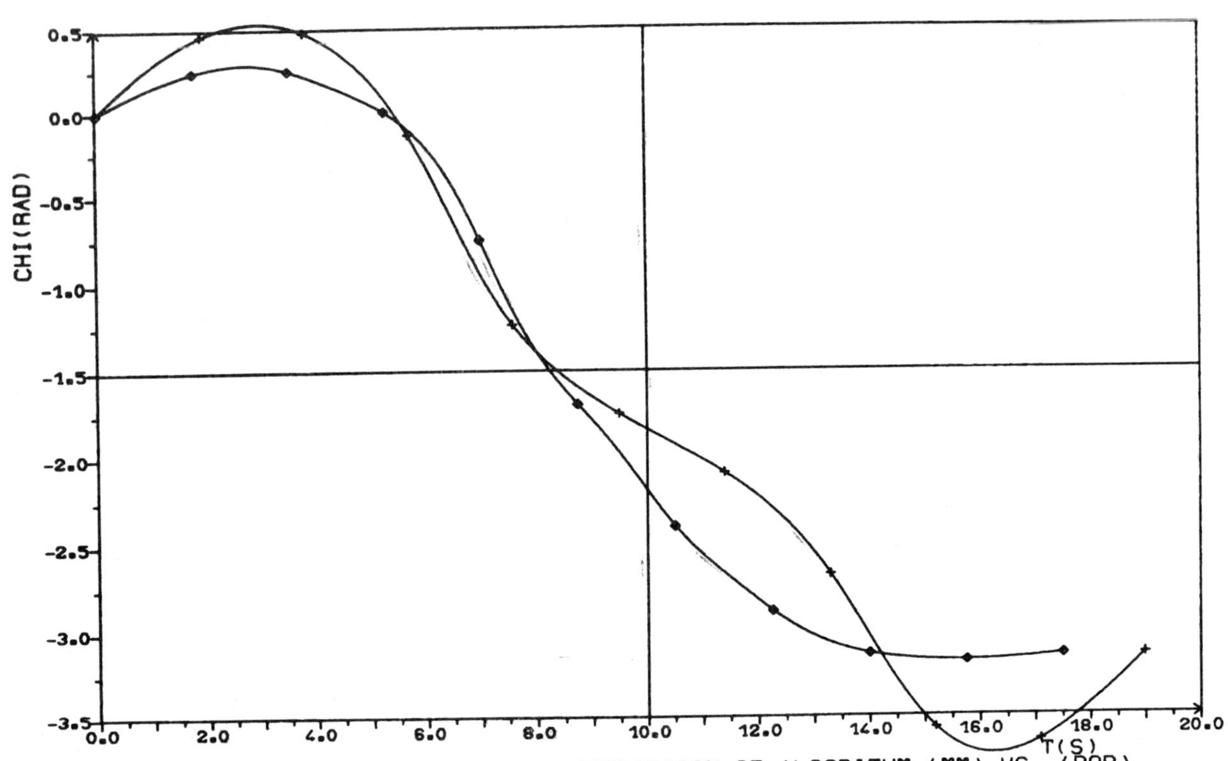

FIG. 5. TIME OPTIMAL TURNING FLIGHT - COMPARISON OF ALGORITHM (MM) VS. (RQP)

GRAPH OF FLIGHTPATH YAW ANGLE

+- ALFA MAX = 30 ◇- ALFA MAX = 90

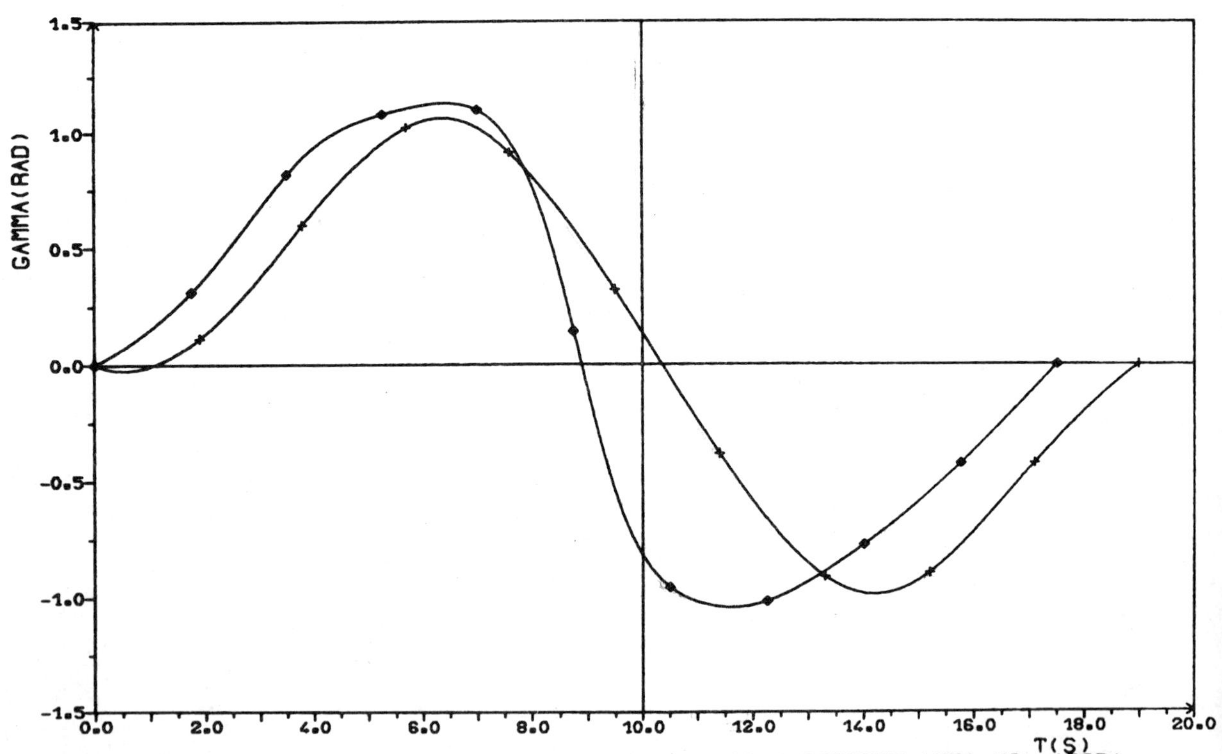

FIG. 6. TIME OPTIMAL TURNING FLIGHT - COMPARISON OF ALGORITHM (MM) VS. (RQP)

GRAPH OF FLIGHTPATH PITCH ANGLE

+- ALFA MAX = 30 ◇- ALFA MAX = 90

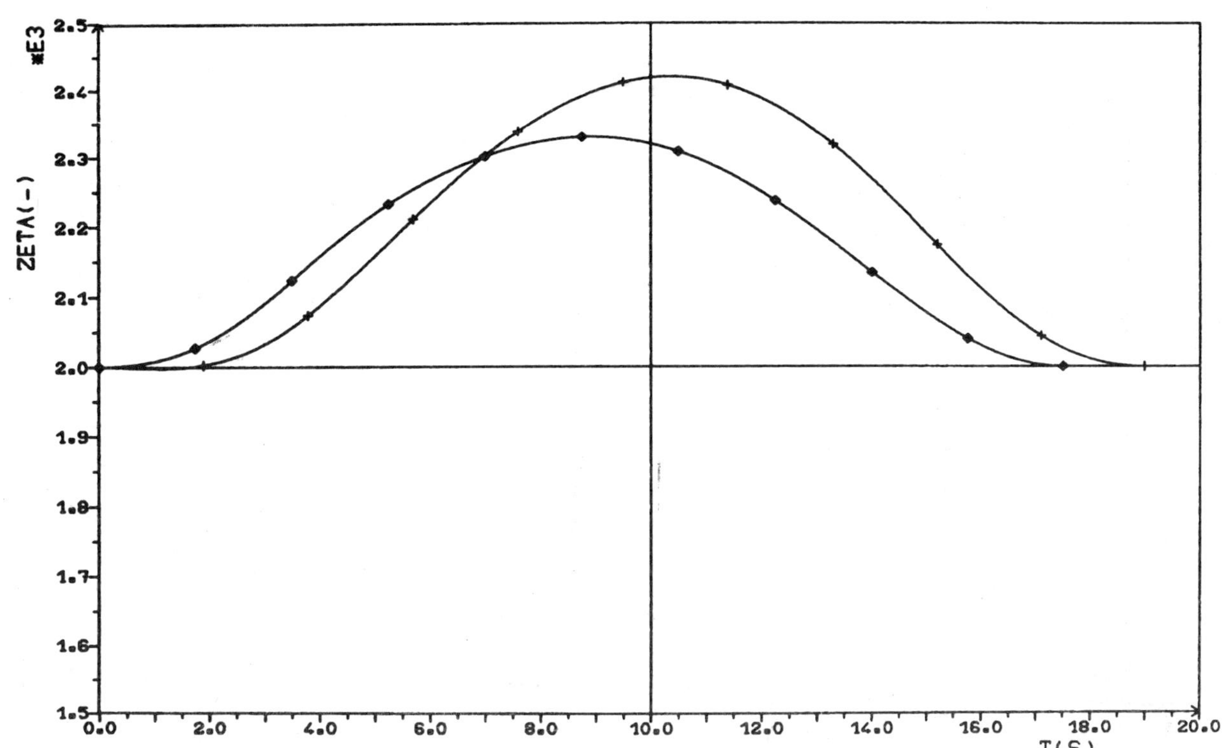

FIG. 7. TIME OPTIMAL TURNING FLIGHT - COMPARISON OF ALGORITHM (MM) VS. (RQP)

GRAPH OF FLIGHT ALTITUDE

+ - ALFA MAX = 30 ◇ - ALFA MAX = 90

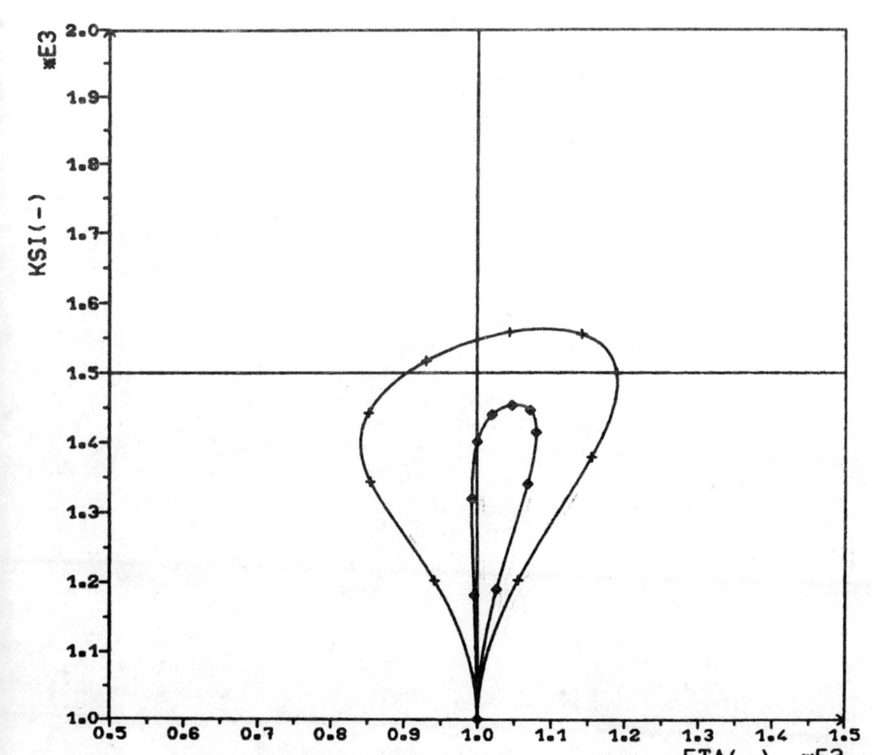

FIG. 8. TIME OPTIMAL TURNING FLIGHT - COMPARISON OF ALGORITHM (MM) VS. (RQP)

PROJECTED TRAJECTORY

+ - ALFA MAX = 30 ◇ - ALFA MAX = 90

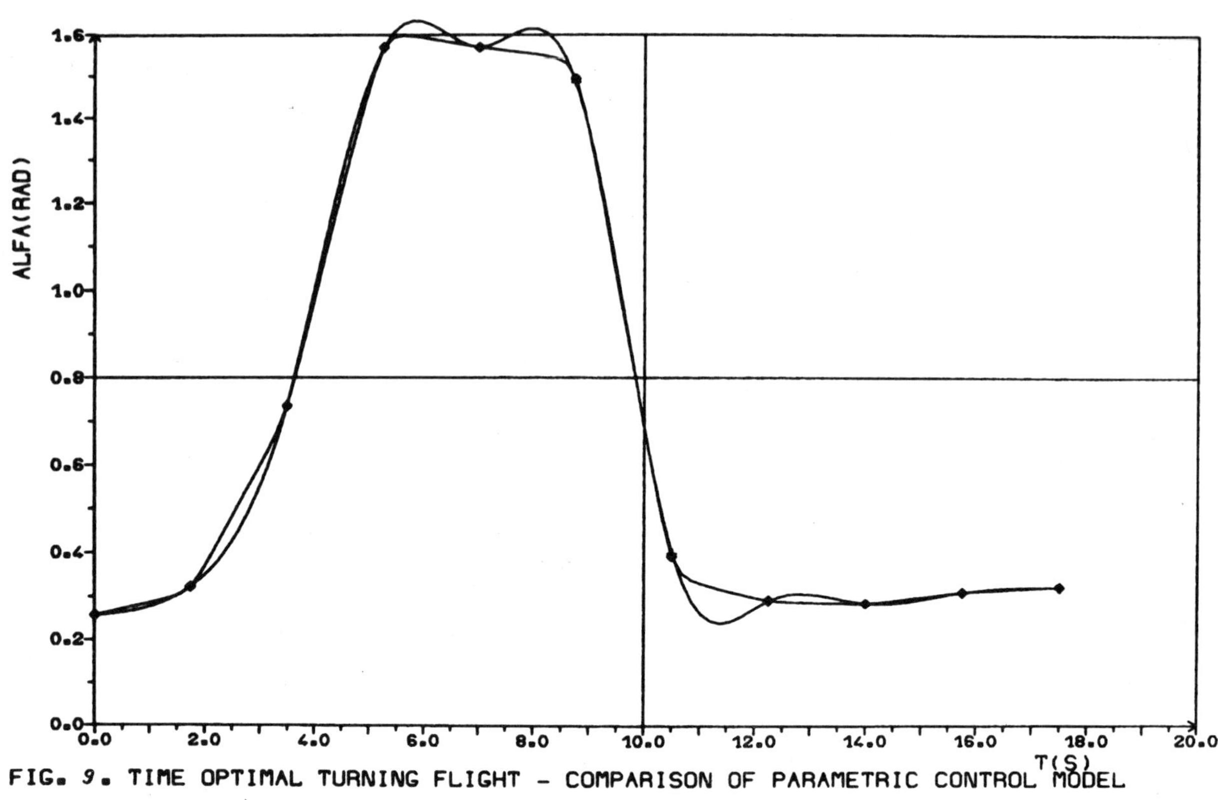

FIG. 9. TIME OPTIMAL TURNING FLIGHT - COMPARISON OF PARAMETRIC CONTROL MODEL

GRAPH OF ANGLE OF ATTACK

+- ALFA MAX = 90 ◇- ALFA MAX = 90 EX

COMPUTER AIDED DESIGN VIA OPTIMIZATION

D. Q. Mayne*, E. Polak and A. Sangiovanni-Vincentelli****

**Department of Computing and Control, Imperial College, London SW7 2BZ, UK*
***Department of Electrical Engineering and Computer Sciences and the Electronics Research Laboratory, University of California, Berkeley, California 94720, USA*

Abstract. Many design problems, including control design problems, involve infinite dimensional constraints of the form $\phi(z,\alpha) \leq 0$ for all $\alpha \in \mathcal{A}$, where α denotes time or frequency or a parameter vector. In other design problems, tuning or trimming of certain parameters, after manufacture of the system, is permitted; the corresponding constraint is that for each α in $\mathcal{A}$ there exists a value τ (of the tuning parameter) in a permissible set T such that $\phi(z,\alpha,t) < 0$. New algorithms for solving design problems having such constraints are described.

Keywords. Computer-aided design, optimization, infinite dimensional constraint, outer approximations, nondifferentiable optimization, control system design.

INTRODUCTION

Infinite dimensional constraints, of the form $\phi(z,\alpha) \leq 0$ for all $\alpha \in \mathcal{A}$, $\mathcal{A}$ a subset of $\mathbb{R}^m$, arise in surprisingly many design problems. Some examples follow:

(i) Design of Envelope-Constrained Filters

The problem here is the choice of weighting function w of a digital filter to process a given input pulse s corrupted by noise such that the output error is minimized subject to the constraint that the noiseless output pulse $\psi = g*s$ satisfies an envelope constraint ($\psi(t) \in [a(t),b(t)]$ for all $t \in [0,t_1]$). The problem is relevant to pulse compression in radar systems, waveform equalization, channel equalization for communication and deconvolution of seismic and medical ultrasonic data.

(ii) Design of Controllers (Zakian and Al-Naib, 1973)

The parameters (z) of a controller are to be chosen so that, inter alia, the resultant closed loop system satisfies certain constraints. These constraints often include hard constraints on controls and states; a typical constraint is $y(z,t) \leq M$ for all $t \in [t_1,t_2]$, where $y(z,\cdot)$ is the response of the closed loop system to a step input. Similarly, frequency domain constraints on loop gain, interaction, stability, may be expressed as $\phi(z,\omega) \leq 0$ for all $\omega \in \Omega$.

(iii) Design of Earthquake Resistant Structures

The problem here is to design structures, such as steel-framed multistory buildings that can resist earthquakes. The design considerations include the constraint that the displacements of structural elements, in response to a specified input, should be limited in magnitude at all times in a certain interval.

An additional range of problems occur when the parameter values of the actual system, structure or device differ from the nominal values employed in the design. This difference may occur because of production tolerances employed in manufacture (e.g. in strcture and circuit design) or because of lack of precise knowledge of some parameters in a system (e.g. identification error). A satisfactory design may require satisfaction of certain constraints not only by the nominal design but also by all possible realizations of the system as the appropriate system parameters range over the tolerance set. Examples include:

(iv) Optimum Design of Chemical Plant with Uncertain Parameters

Certain design constraints must be satisfied for all values of the uncertain parameters in a linear set.

(v) Design of Robust Controllers

The controller must be such that the design constraints are satisfied for all values of certain plant parameters lying in a specified set.

(vi) Circuit Design

Design constraints must be met not only by the nominal design $\hat{z} \in \mathbb{R}^p$ but for all values of the circuit parameter z in the set $\hat{z} + \mathcal{A}$, where $\mathcal{A}$ is the tolerance set. The set is commonly a hypercube.

The latter set of constraints ($\max\{\phi(z,\alpha)|\alpha \in \mathcal{A}\} \leq 0$) may be very difficult to satisfy and may therefore require, if a 100% yield in manufacture is required, very tight tolerances (i.e. a "small" $\mathcal{A}$), making manufacture prohibitively expensive. To avoid this difficulty the facility for altering certain parameters ("tuning" controllers, "trimming" circuit components) after manufacture is often provided. If tuning or trimming is effected by a parameter τ ranging over a compact set T, then the constaint has the form: for each $\alpha \in \mathcal{A}$ there exists a $\tau \in T$ such that $\xi(z,\alpha,\tau) \leq 0$, or, equivalently:

$$\max_{\alpha \in \mathcal{A}} \min_{\tau \in T} \xi(z,\alpha,\tau) \leq 0$$

Taking into account conventional (finite dimensional) constaints, many design problems may be expressed either as:

A. Determine a $z \in F$.

or:

B. Minimize $\{f(z)|z \in F\}$

where:

$$F \triangleq \{z \in \mathbb{R}^p \,|\, g(z) \leq 0,\ \psi_{\mathcal{A}}(z) \leq 0\} \qquad (1)$$

where $g : \mathbb{R}^p \to \mathbb{R}^q$, and $\psi_{\mathcal{A}} : \mathbb{R}^p \to \mathbb{R}$; $\psi_{\mathcal{A}}$ is defined by:

$$\psi_{\mathcal{A}}(z) = \max_{\alpha \in \mathcal{A}} \max_{j \in \underline{r}} \phi^j(z,\alpha) \qquad (2)$$

where $\underline{r}$ denotes the set $\{1,2,\ldots,r\}$. If post-manufacture tuning or trimming is permitted, the design problems are:

C. Determine a $z \in F$.

D. Minimize $\{f(z)|z \in F\}$

where F is now defined by:

$$f \triangleq \{z \in \mathbb{R}^p \,|\, g(z) \leq 0,\ \psi_{\mathcal{A},T}(z) \leq 0\} \qquad (3)$$

and $\psi_{\mathcal{A},T} : \mathbb{R}^p \to \mathbb{R}$ is defined by:

$$\psi_{\mathcal{A},T}(z) \triangleq \max_{\alpha \in \mathcal{A}} \min_{\tau \in T} \max_{j \in \underline{r}} \xi^j(z,\alpha,\tau) \qquad (4)$$

These problems are obviously very complex — merely to test feasibility requires a global solution of a maximization problem (see (2)) for problems A and B and of a max-min problem (see (4)) for problems C and D. Conditions of optimality for B have been derived and conceptual algorithms, developed by, for example, Demyanov (1966); these algorithms require the solution of infinite dimensional linear programs at each iterations. Another interesting class of conceptual algorithms are the outer approximations algorithms of Levitin and Polyak (1966) Eaves and Zangwill (1971) and Blankenship and Falk (1974). Indeed the only implementable algorithms (i.e. requiring only a finite number of operations at each iteration) appear to be those of the authors and their collaborators. This paper presents the essential features of these new algorithms so that their utility for design (especially control design) may be assessed.

ALGORITHMS FOR DESIGN PROBLEMS WITH INFINITE DIMENSIONAL CONSTRAINTS

The master algorithms for this type of problem (see (i)-(vii) above) may be divided into four classes:

1. Solving A using implementable "feasible directions" algorithms.
2. Solving A using implementable outer approximations algorithms.
3. Solving B using implementable "feasible directions" algorithms.
4. Solving B using implementable outer approximations algorithms.

The essential features of the master algorithms are retained if we ignore the conventional constraints and restrict the number of infinite dimensional constraints to one so that F is defined by:

$$F \triangleq \{z \in \mathbb{R}^p \,|\, \psi_{\mathcal{A}}(z) \leq 0\} \qquad (5)$$

where, now, $\psi_{\mathcal{A}} : \mathbb{R}^p \to \mathbb{R}$ is defined by:

$$\psi_{\mathcal{A}}(z) \triangleq \max_{\alpha \in \mathcal{A}} \phi(z,\alpha) \qquad (6)$$

2.1 Feasible Directions Type Algorithms for Problems A and B

We assume that $\phi : \mathbb{R}^p \times \mathcal{A} \to \mathbb{R}$ is continuously differentiable, that $\mathcal{A}$ is compact subset of $\mathbb{R}$ and that, for each z, $\phi(z,\cdot)$ has only a finite number of local maxima in $\mathcal{A}$.

If z is not feasible ($\psi_{\mathcal{A}}(z) > 0$) the feasible directions" algorithms for Problem A determine a search direction which is a descent direction for $\psi_{\mathcal{A}}(z)$. For any z in $\mathbb{R}^p$ let the "ε-most-active constraint" set $\mathcal{A}_\varepsilon(z) \subset \mathcal{A}$ be defined by:

$$\mathcal{A}_\varepsilon(z) \triangleq \{\alpha \in \mathcal{A} | \phi(z,\alpha) \geq \psi_{\mathcal{A}}(z) - \varepsilon\} \qquad (7)$$

A solution $h_\varepsilon(z)$ of:

$$\theta_\varepsilon(z) = \min_{h \in S} \max_{\alpha \in \mathcal{A}_\varepsilon(z)} \langle \nabla_z \phi(z,\alpha), h \rangle \qquad (8)$$

(where S is the unit hypercube in $\mathbb{R}^p$) is a descent direction for $\psi_{\mathcal{A}}(z)$ <u>if</u> $\theta_\varepsilon(z) < 0$; solving (8) is prohibitively difficult. For an implementable algorithm $\mathcal{A}_\varepsilon(z)$ must be replaced (approximated) by a finite set. One such approximation was employed in Polak and Mayne (1976); a better approximation, $\mathcal{A}_\varepsilon(z)$, proposed in Gonzaga and Polak (1979), is:

$$\mathcal{A}_\varepsilon(z) \triangleq \{\alpha \in \mathcal{A} | \alpha \text{ is a local maximizer of } \phi(z,\cdot)\} \qquad (9)$$

The search direction $\tilde{h}_\varepsilon(z)$ is any solution of the finite-dimensional linear program:

$$\tilde{\theta}_\varepsilon(z) \triangleq \min_{h \in S} \max_{\alpha \in \mathcal{A}_\varepsilon(z)} \langle \nabla_z \phi(z,\alpha), h \rangle \quad (10)$$

To ensure the existence of a descent direction for $\psi_{\mathcal{A}}(z)$ we assume that the set $\{\nabla_z \phi(z,\alpha),\ \alpha \in \mathcal{A}_0(z)\}$ is positive linearly independent.

Algorithm 1 (to compute a $z \in F$)

Data: $z_0 \in \mathbb{R}^p$, $\varepsilon_0 \in (0,\infty)$, $\beta \in (0,1)$.

Step 0: Set $i = 0$, set $\varepsilon = \varepsilon_0$.

Step 1: Compute $\tilde{\mathcal{A}}_\varepsilon(z_i)$, $\tilde{\theta}_\varepsilon(z_i)$, $\tilde{h}_\varepsilon(z_i)$

Step 2: If $\tilde{\theta}_\varepsilon(z_i) > -\varepsilon$, set $\varepsilon = \varepsilon/2$ and go to Step 1.

Step 3: If $z \in F$ stop. Else compute the largest $\lambda_i \in \{1,\beta,\beta^2,\ldots\}$ such that

$$\psi_{\mathcal{A}}(z_i + \lambda_i \tilde{h}_\varepsilon(z_i)) - \psi_{\mathcal{A}}(z_i) \le -\lambda_i\ \varepsilon/2$$

Step 4: Set $i = i+1$ and go to Step 1. ¤

It can be shown that this algorithm finds a feasible point in a finite number of iterations. The implementable version (Mayne and Polak, 1976; Gonzaga and Polak, 1979) approximates $\mathcal{A}$ by a discretization $\hat{\mathcal{A}}_\delta$ (e.g. of $\mathcal{A} = [0,1]$, $\hat{\mathcal{A}}_\delta = \{0,\delta,2\delta,\ldots,1\}$ and refines this discretization automatically via an adaptive law which ensures convergence; thus $\mathcal{A}_\varepsilon(z_i)$ is replaced by an easily determined approximation.

For problem B ($\min\{f(z) \mid z \in F\}$) we make the extra assumption that f is continuously differentiable and compute the search direction $\tilde{h}_\varepsilon(z)$ to be the solution of:

$$\tilde{\theta}_\varepsilon(z) \triangleq \min_{h \in S} \max\{\langle \nabla f(z), h \rangle - \gamma \psi_{\mathcal{A}}(z)_+;\ \langle \nabla_z \phi(z,\alpha), h \rangle,\ \alpha \in \tilde{\mathcal{A}}_\varepsilon(z)\} \quad (11)$$

where γ is a positive constant and $\psi_{\mathcal{A}}(z)_+ \triangleq \max\{0, \psi_{\mathcal{A}}(z)\}$. We note that $\tilde{\theta}_\varepsilon(z) \le 0$ and that if $\tilde{\theta}_\varepsilon(z) < 0$ then:

$$\langle \nabla f(z), \tilde{h}_\varepsilon(z) \rangle < \gamma \psi_{\mathcal{A}}(z)_+ \ \langle \nabla_z \phi(z,\alpha), \hat{h}_\varepsilon(z) \rangle < 0$$

for all $\alpha \in \tilde{\mathcal{A}}_\varepsilon(z)$.

Hence $\tilde{h}_\varepsilon(z)$ is a descent direction for $\phi(z,\alpha)$ at all $\alpha \in \tilde{\mathcal{A}}_\varepsilon(z)$ and also for $f(z)$ if $z \in F$; if $z \notin F$, $\tilde{h}_\varepsilon(z)$ permits an increase in f. This permitted increase decreases as z approaches F ($\psi_{\mathcal{A}}(z)_+ \to 0$).

Algorithm 2, a "feasible-directions" algorithm for Problem B, is similar to Algorithm 1, except that $\tilde{h}_\varepsilon(z)$ is computed from (11) and Step 3 is replaced by:

Step 3': If $\psi_{\mathcal{A}}(z_i) > 0$, compute the largest $\lambda_i \in \{1,\beta,\beta^2,\ldots\}$ such that:

$$\psi_{\mathcal{A}}(z_i + \lambda_i \tilde{h}_\varepsilon(z_i)) - \psi_{\mathcal{A}}(z_i) \le -\lambda_i\ \varepsilon/2$$

If $\psi_{\mathcal{A}}(z_i) \le 0$, compute the largest $\lambda_i \in \{1,\beta,\beta^2,\ldots\}$ such that:

$$f(z_i + \lambda_i h\ (z_i)) - f(z_i) \le -\lambda_i\ \varepsilon/2;\ \psi_{\mathcal{A}}(z_i) \le 0$$

¤

In the implementable version of the algorithm $\mathcal{A}$ is again replaced by a discretization which is automatically refined to ensure convergence (Polak and Mayne, 1976; Gonzaga and Polak, 1979).

2.2 Outer Approximation Algorithms for Problems A and B

We assume again that ϕ and f are continuously differentiable and that $\mathcal{A}$ is a compact subset of $\mathbb{R}^s$. For any subset $\mathcal{A}'$ of $\mathbb{R}^s$ let $F_{\mathcal{A}'}$ be defined by:

$$F_{\mathcal{A}'} = \{z \in \mathbb{R}^p \mid \phi(z,\alpha) \le 0 \text{ for } \alpha \in \mathcal{A}'\}$$

If $\mathcal{A}' \subset \mathcal{A}$, then:

(i) $F_{\mathcal{A}} \subset F_{\mathcal{A}'}$

($F_{\mathcal{A}'}$ is an outer approximation of $F_{\mathcal{A}}$) and:

(ii) $\min\{f(z) \mid z \in F_{\mathcal{A}'}\} \le \min\{f(z) \mid z \in \mathcal{A}\}$

The outer approximation algorithms (Eaves and Zangwill, 1971; Blankenship and Falk 1974) employ a sequence of outer approximations, $\{F_{\mathcal{A}_i}\}$ as in the following conceptual algorithm for Problem A.

Algorithm 3 (to determine $z \in F$).

Data: $\mathcal{A}_0$ (a discrete subset of $\mathcal{A}$).

Step 0: Set $i = 0$.

Step 1: Compute a z_i in $F_{\mathcal{A}_i}$.

Step 2: Compute $\psi_{\mathcal{A}}(z_i)$ and a $\alpha_i \in \mathcal{A}$ such that $\phi(z_i,\alpha_i) = \psi_{\mathcal{A}}(z_i)$. If $\psi_{\mathcal{A}}(z_i) \le 0$, stop.

Step 3: Set $\mathcal{A}_{i+1} = \mathcal{A}_i \cup \{\alpha_i\}$. Set $i = i+1$ and go to Step 1. ¤

It can be shown (Eaves and Zangwill, 1971; Blankenship and Falk, 1974) that any accumulation point of an infinite sequence generated by Algorithm 3 is feasible. Step 1 can be achieved in a finite number of iterations (Mayne and Colleagues, 1979 Mayne and Polak, 1979), but step 2 involves an infinite process. Moreover the cardinality of $\mathcal{A}_i$ tends to infinity with i. To cope with the latter difficulty a method for dropping constraints is required. Several schemes for achieving this are proposed in (Mayne and Colleagues, 1979; Gonzaga and Polak, 1979). The next algorithm incorporates a scheme (see Step 3) which has proved successful (Gonzaga and Polak, 1979).

Algorithm 4 (to determine a $z \in F$)

Data: $\mathcal{A}_0$ (a discrete subset of $\mathcal{A}$),
$\delta \in (0,1)$, $k \gg 1$.

Step 0: Set $i = 0$.

Step 1: Compute a z_i in $F_{\mathcal{A}_i}$.

Step 2: Compute $\psi_{\mathcal{A}}(z_i)$ and a $\alpha_i \in \mathcal{A}$ such that $\phi(z_i,\alpha_i) = \psi_{\mathcal{A}}(z_i)$. If $\psi_{\mathcal{A}}(z_i) \leq 0$, stop.

Step 3: Set $\mathcal{A}_{i+1} = \{\alpha_j | \phi(z_j,\alpha_j) \geq k(\delta^j - \delta^i),\ j = 0,1,2,\ldots,i\}$ (12)

Set $i = i + 1$ and go to Step 1. ¤

It follows from (Gonzaga and Polak, 1979) that any accumulation point $\hat{z}$ of an infinite sequence $\{z_i\}$ generated by algorithm 4 is feasible. Note that for any j, $k(\delta^j - \delta^i) = 0$ when $i = j$ and $k(\delta^j - \delta^i) \to k\delta^i$ as $i \to \infty$; hence the test in Step 3 increases in difficulty with i so that α_j will probably be dropped from $\mathcal{A}_i$ for all i sufficiently large, thus controlling the growth of $\mathcal{A}_i$. The implementable version of this algorithm utilizes an approximate, but progressively more precise computation of α_i (recall α_i solves $\max\{\phi(z_i,\alpha) | \alpha \in \mathcal{A}\}$).

A conceptual algorithm (algorithm 5) for solving Problem B is obtained (Eaves and Zangwill, 1971; Blankenship and Falk, 1974) if Step 1 in algorithm 4 is replaced by:

Step 1': Compute a z_i to solve:

$$P_{\mathcal{A}_i} : \min\{f(z) | z \in \mathcal{A}_i\}$$

Clearly Step 1' must be replaced by an approximate solution to $P_{\mathcal{A}_i}$ in an implementable algorithm. This requires an optimality function to gauge the accuracy of the solution to $P_{\mathcal{A}_i}$. A suitable function is $\theta_{\mathcal{A}'}$ defined by:

$$\theta_{\mathcal{A}'}(z) = \min_{h \in S} \max\{\langle \nabla f(z),h \rangle ;\ \phi(z,\alpha) + \langle \nabla_z \phi(z,\alpha),h \rangle ,\alpha \in \mathcal{A}'\} - \psi_{\mathcal{A}}(z)_+ \} \quad (13)$$

Under weak conditions (of continuous differentiability of f and ϕ and positive linear independence of the most active constraints) it can be shown that:

(i) $\theta_{\mathcal{A}'}(z) < 0$ for all $z \notin F_{\mathcal{A}'}$ (14)

(ii) $\theta_{\mathcal{A}'}(z) < 0$ for all $z \in \mathbb{R}^p$ (15)

(iii) $\theta_{\mathcal{A}'}(z) = 0$ if and only if $z \in F$,

and satisfies the F. John optimality condition for the problem $P_{\mathcal{A}'}$: $\min\{f(z) | z \in F_{\mathcal{A}'}\}$. The conceptual step 1' in algorithm 5 can now be replaced, yielding:

Algorithm 6 (to solve $\min\{f(z) | z \in F\}$)

Data: $\mathcal{A}_0$ (a discrete subset of $\mathcal{A}$),
γ, $\delta \in (0,1)$, $k \gg 1$.

Step 0: Set $i = 0$.

Step 1: Solve $P_{\mathcal{A}_i}$ approximately to obtain a z_i such that:

$$\theta_{\mathcal{A}_i}(z_i) \geq -\gamma^i.$$

Step 2: Compute $\psi_{\mathcal{A}}(z_i)$ and $\alpha_i \in \mathcal{A}$ satisfying $\psi_{\mathcal{A}}(z_i) = \phi(z_i,\alpha_i)$.

Step 3: Set $\mathcal{A}_{i+1} = \{\alpha_j | \phi(z_j,\alpha_j) \geq k(\delta^j - \delta^i),\ j = 0,1,2,\ldots,i\}$.

Set $i = i + 1$ and go to Step 1. ¤

It can be shown (Gonzaga and Polak, 1979), that any accumulation point $\hat{z}$ of an infinite sequence $\{z_i\}$ generated by algorithm 6 is feasible and satisfies $\theta_{\mathcal{A}}(\hat{z}) = 0$, a necessary condition of optimality for $P_{\mathcal{A}}$. As before, the implementable version of this algorithm solves $\max\{\phi(z_i,\alpha) | \alpha \in \mathcal{A}_i\}$ (in Step 2) approximately but with increasing precision as i increases.

2.3 Subalgorithms for the Master Algorithms

The following sub-algorithms are required by the master algorithms described above

(i) Standard linear programs (e.g. in Step 1 of Algorithms 1 and 2).

(ii) Algorithms for solving a finite number of inequalities in a finite number of iterations (e.g. in Step 1 of Algorithms 3 and 4). Two new algorithms (Mayne and Colleagues, 1979a; Polak and Mayne, 1978a) have been developed for this purpose. These algorithms combine the quadratic rate of convergence of Newton's method with the robustness and finite convergence of first order methods, and have proved particularly successful within the outer approximation master algorithms since they generate a point in the interior of the (current) constraint set.

(iii) Algorithms for Constrained Optimization (e.g. in Step 1 of Algorithms 5 and 6). Two new algorithms (Mayne and Polak, 1978; Polak and Mayne 1978b) which are globally stabilized versions of Newton's method (using, respectively, an exact penalty function or hybriding with a phase I - phase II method of feasible directions (Polak and Colleagues, 1979) to achieve stabilization) have been developed.

Algorithms 2 and 3 employ (see Step 1) a special Phase I - Phase II method of feasible directions (Polak and Mayne, 1976; Gonzaga and Colleagues, 1978; Polak and Colleagues, 1979).

ALGORITHMS FOR DESIGN PROBLEMS WITH INFINITE DIMENSIONAL CONSTRAINTS AND TUNING

Again for exposition we consider the simplest case of one infinite dimensional constraint with tuning so that the feasible set F is defined by:

$$F = \{z \in \mathbb{R}^p \mid \psi_{\mathcal{A},T}(z) \leq 0\} \tag{16}$$

where, now, $\psi_{\mathcal{A},T} : \mathbb{R}^p \to \mathbb{R}$ is defined by:

$$\psi_{\mathcal{A},T}(z) \triangleq \max_{\alpha \in \mathcal{A}} \min_{\tau \in T} \xi(z,\alpha,\tau) \tag{17}$$

It is assumed that f and ξ are continuously differentiable, that $\mathcal{A}$ is a compact subset of $\mathbb{R}^s$ and T is a compact subset of $\mathbb{R}^v$. We will consider only Problem D: $\min\{f(z) \mid z \in F\}$.

3.1 A Generalized Gradient Algorithm

Since F is defined by (16) and (17), it is clear that any implementable algorithm will have to approximate the (infinite) set $\mathcal{A}$ by a set of discrete approximations $\{\mathcal{A}_i\}$. The next algorithm (algorithm 7) is a conceptual algorithm based in outer approximations, for solving $\min\{f(z) \mid z \in F\}$:

Algorithm 7 (to solve $P_{\mathcal{A},T} : \min\{f(z) \mid \psi_{\mathcal{A},T}(z) \leq 0\}$

Data: $\mathcal{A}_0$ (a discrete subset of $\mathcal{A}$), $\gamma, \delta \in (0,1), k >> 1$.

Step 0: Set $i = 0$.

Step 1: Solve $P_{\mathcal{A}_i,T} : \min\{f(z) \mid \psi_{\mathcal{A}_i,T}(z) \leq 0\}$ for z_i.

Step 2: Compute $\psi_{\mathcal{A},T}(z_i)$ and $\alpha_i \in \mathcal{A}$ satisfying $\xi(z_i,\alpha_i,\tau) = \psi_{\mathcal{A},T}(z_i)$ for some $\tau \in T$.

Step 3: Set

$$\mathcal{A}_{i+1} = \{\alpha_j \mid \psi_{\mathcal{A},T}(z_j) \geq k(\delta^j - \delta^j), \quad j = 0,1,2,\ldots,i\} \tag{18}$$

Set $i = i + 1$ and go to Step 1. ¤

This algorithm can be seen to involve a small, and obvious, modification, in Step 2, to (the conceptual) algorithm 5, and reduces $P_{\mathcal{A},T}$ to an infinite sequence of problems $\{P_{\mathcal{A}_i,T}\}$ where $\mathcal{A}_i$ is a discrete set. However solving $P_{\mathcal{A}_i,T}$ is not trivial since the function $\psi \triangleq \psi_{\mathcal{A}_i,T}$ is, in general, not differentiable, and may even fail to have directional derivatives. However, ψ_i is locally Lipschitz, and, hence has a generalized gradient (Clarke, 1975) $\partial\psi_i(z)$ at z ($\partial\psi(z)$) is defined to be the convex hull of the set of all limits of the form $\lim_{j\to\infty} \nabla\psi(z+v_j)$ where $v_j \to 0$ as $j \to \infty$ in such a way that $\nabla\psi(z+v_j)$ is well defined; thus the generalized gradient $\partial\psi$ of $\psi \triangleq \max\{h^1,h^2\}$, where h^1 and h^2 are continuously differentiable, satisfies $\partial\psi(z) = \{\nabla h^1(z)\}$ if $h^1(z) > h^2(z)$, $\partial\psi(z) = \{\nabla h^2(z)\}$ if $h^2(z) > h^1(z)$ and $\partial\psi(z) = co\{\nabla h^1(z), \nabla h^2(z)\}$ if $h^1(z) = h^2(z)$). Clearly $\partial\psi_i$ is a point-to-set map, mapping $\mathbb{R}^p$ into $2^{\mathbb{R}^r}$. It can be shown (Clarke, 1979) for all z in $\mathbb{R}^r$, that $\partial\psi(z)$ is a well defined non-empty subset of $\mathbb{R}^p$ and that the map $\partial\psi$ is bounded and upper semi-continuous on any open bounded subset of $\mathbb{R}^p$. A necessary condition (Clarke, 1979) for $\hat{z}$ to be a solution of $\min\{\psi(z)\}$ is that $0 \in \partial\psi(z)$. The steepest descent direction for ψ at z is $-Nr\{\partial\psi(z)\}$, where

$$Nr\{\partial\psi(z)\} \triangleq \arg\min\{\|h\| \mid h \in \partial\psi(z)\} \tag{19}$$

However this direction cannot be used in an algorithm to reduce ψ because discontinuities in $\partial\psi$ may cause jamming (e.g. consider an algorithm to find a z satisfying $\psi(z) \leq 0$ where $\psi \triangleq \max\{h^1,h^2\}$). As in Algorithm 1 (see eq. (8)) some local averaging is required. For this purpose a smeared generalized gradient (Goldstein, 1977; Polak and Sangiovanni-Vincentelli, 1979) $\partial_\varepsilon\psi : \mathbb{R}^r \to 2^{\mathbb{R}^r}$, defined as follows:

$$\partial_\varepsilon\psi(z) \triangleq co \bigcup_{y \in B(z,\varepsilon)} \partial\psi(y) \tag{20}$$

is employed ($B(z,\varepsilon) \triangleq \{y \mid \|y-z\| \leq \varepsilon\}$). This smeared generalized gradient can be employed to solve $P_{\mathcal{A}_i,T}$ in an extension of a feasible-directions type algorithm. For any $\varepsilon > 0$ we define the set $M_\varepsilon(z)$ by

$$M_\varepsilon(z) \triangleq \begin{cases} \{\nabla f(z)\} & \text{if } \psi(z) < \varepsilon \\ co\{\{\nabla f(z)\} \cup \partial\psi(z)\} & \text{if } \psi(z) \geq \varepsilon \end{cases} \tag{21}$$

(recall that $\psi \triangleq \psi_{\mathcal{A}_i,T}$). The optimality function $\theta_\varepsilon : \mathbb{R}^p \to \mathbb{R}$ for $P_{\mathcal{A}_i,T}$ is defined by:

$$\theta_\varepsilon(z) \triangleq -\min\{\|h\|^2 \mid h \in M_\varepsilon(z)\} \tag{22}$$

and the descent direction $h_\varepsilon(z)$ by:

$$h_\varepsilon(z) \triangleq -Nr\{M_\varepsilon(z)\} \tag{23}$$

The algorithm for solving $P_{\mathcal{A}_i,T}$ is:

Algorithm 8 (to solve $P_{\mathcal{A}_i,T}$) (Polak and Sangiovanni-Vincentelli, 1979).

Data: $z_0 \in F_i \triangleq \{z|\psi(z) \leq 0\}$, $\varepsilon_0 > 0$, $\beta \in (0,1)$.

Step 0: Set $j = 0$.

Step 1: Set $\varepsilon = \varepsilon_0$.

Step 2: Compute $h_\varepsilon(z_j)$ and $\theta_\varepsilon(z_j)$.

Step 3: If $\theta_\varepsilon(z_j) \geq -\varepsilon$, set $\varepsilon = \varepsilon/2$ and go to Step 2.

Step 4: If $\theta_0(z_j) = 0$ stop. Else compute the largest $\lambda_j \in \{1,\beta,\beta^2,\ldots\}$ such that

$$f(z_j + \lambda_j h_\varepsilon(z_j)) - f(z_j) \leq -\lambda_i \, \varepsilon/2 \text{ and}$$

$$\psi(z_j + \lambda_j h_\varepsilon(z_j)) \leq 0.$$

Step 5: Set $z_{j+1} = z_j + \lambda_j h_\varepsilon(z_j)$, set $j = j + 1$ and go to step 1. ⌑

This algorithm can also be used to find $z_0 \in F_i$ or a combined phase 1 - phase 2 algorithm can be constructed, on the lines of (Polak and Colleagues, 1979), so that z_0 may be any point in $\mathbb{R}^p$. The resultant algorithm may then be employed in algorithm 7. If Step 1 is replaced by:

Step 1': Solve $P_{\mathcal{A}_i,T}$ approximately to obtain a z_i such that $\theta_0(z_i) \geq -\gamma_i$

then a more practical algorithm is obtained; this step may be implemented using a finite number of iterations of algorithm 8. However, the algorithm is still not implementable because $\partial_\varepsilon\psi(z_j)$ is an infinite set which requires an infinite process to compute. Hence the final, implementable algorithm, described by Polak and Sangiovanni-Vincentelli (1979), makes use of the fact that $\psi \triangleq \psi_{\mathcal{A}_i,T}$ is semi-smooth to approximate $\partial\psi$ by a finite number of vector; the precision of the approximation is adaptively increased in such a way that convergence is ensured; it also employs an approximate, but progressively more accurate, estimation of the α_i required in Step 2 of Algorithm 7.

3.2 A Transformation Approach

The algorithm described in 3.1 suffers two disadvantages: the first, common to most non-differentiable optimization procedures is that the bisection procedure to obtain an approximation to the generalized gradient is computationally very expensive; the second is that it is limited to the case when there is only one constraint of the form $\max_{\alpha\in\mathcal{A}} \min_{\tau\in T} \xi(z,\alpha,t) \leq 0$ (i.e. $\underline{r} = 1$).

The latter restriction arises from the requirement of semismoothness of $\partial\psi_{\mathcal{A}_i,T}$.

Very recently a method (Polak, 1979) of avoiding these difficulties has been found.

As in Section 3.1 a master algorithm (e.g. algorithm (7) with Step 1 replaced, as before, by Step 1' is employed; the difference lies in the sub-algorithm to solve $P_{\mathcal{A}_i,T}$. Instead of solving $P_{\mathcal{A}_i},T$, directly, the problem is transformed into an equivalent problem soluble by conventional optimization algorithms. Recall that $P_i \triangleq P_{\mathcal{A}_i,T}$ is defined by:

$$P_i : \min\{f(z)|\psi_{\mathcal{A}_i,T}(z) \leq 0\} \tag{22}$$

where $\psi_{\mathcal{A}_i,T}$ is defined by:

$$\max_{\alpha\in\mathcal{A}_i} \min_{\tau\in T} \xi(z,\alpha,\tau) \leq 0 \tag{23}$$

Suppose that $\mathcal{A}_i = \{\alpha_1,\alpha_2,\ldots,\alpha_J\}$. Then (22) may be written as:

$$P_i : \min\{f(z)|\min_{\tau\in T} \xi(z,\alpha_i,\tau) \leq 0,\ i \in \underline{J}\} \tag{24}$$

Solving P_i requires the determination of a $(\hat{z},\hat{\tau}_1,\hat{\tau}_2,\ldots,\hat{\tau}_J) \in \Gamma \triangleq \mathbb{R}^p \times T \times T \times \ldots \times T$ such that:

$$\xi(\hat{z},\alpha_i,\hat{\tau}_i) \leq 0, \text{ for all } i \in \underline{J} \tag{25}$$

and $f(\hat{z}) \leq f(\tilde{z})$ for any $(\tilde{z},\tau_1,\tau_2,\ldots,\tau_J) \in \Gamma$ satisfying (25). It is therefore plausible (and easily proven) that P_i is equivalent to:

$$\hat{P}_i : \min_{(z,\tau_1,\ldots,\tau_J)} \{f(z)|\xi(z,\alpha_i,\tau_i) \leq 0 \text{ for all } i \in \underline{J}\} \tag{26}$$

But (26) is a conventional constrained optimization problem. The final algorithm is therefore similar to the implementable version of Algorithm 7 but with Algorithm 8 (the subalgorithm required in Step 1') replaced by a conventional algorithm for solving $\tilde{P}_i$. The algorithm is easily extended to deal with ease when $\psi_{\mathcal{A},T}$ is defined by (4) ($\underline{r} > 1$) and z is also subject to conventional and infinite dimensional constraints.

CONTROL DESIGN

The algorithm described above may be employed in a variety of control design problems, in particular Problems (ii), (iv) and (v) of Section 1. The only point requiring further attention is the calculation of derivatives such as $\nabla_z\phi(z,\alpha)$ and $\nabla_z\xi(z,\alpha,\tau)$. If ϕ and ζ represent constraints in the frequency domain, the standard computations only are involved. If, however, they represent constraints in the time domain, the derivatives of the state transition matrix of the closed loop system is repeatedly required. One suggestion for this is outlined by Becker (1979). Preliminary studies of control design using frequency domain (Polak and Mayne, 1976; Gonzaga and Colleagues, 1978; Voreadis, 1978) and time domain (Becker and Colleagues, 1978) constraint have been encouraging.

ACKNOWLEDGEMENT

Research sponsored by the National Science Foundation (RANN) Grant PFR79-08261.

REFERENCES

Becker, R.G., A.L. Heunis and D.Q. Mayne, (1978). Computer aided design of control systems via optimization. Publication 78/47, Department of Computing and Control, Imperial College, London.

Becker, R.G., (1979). Linear system functions via diagonalization. Report, Department of Computing and Control, Imperial College, London.

Blankenship, J.W. and J.E. Falk, (1979). Infinitely constraint optimization problems. The George Washington University, Institute of Management Science and Engineering, Serial T-301.

Clarke, F.M., (1975). Generalized gradients and applications. Trans. Amer. Math. Soc., 205, pp. 247-262.

Demyanov, V.F., (1966). On the solution of certain min max problems. Kybern., 2,47.

Eaves, B.C. and W.I. Zangwill, (1971). Generalized cutting plane algorithms. SIAM Journal of Control, 9, No. 4, 529.

Gonzaga, C., E. Polak and R. Trahan, (1978). An improved algorithm for optimization problems with functional inequality constraints. Report UCB/ERL M78/56, ERL, College of Engineering, University of California, Berkeley.

Gonzaga, C. and E. Polak, (1979). On constraint dropping schemes and optimality functions for a class of outer approximations algorithms. SIAM J. on Control and Optimization, 17, No. 4, 477.

Levitin, E.S. and B.T. Polyak, (1966). Constrained minimization methods. Zn. Vychisl. Mat. Fiz., 6, No. 5, 787.

Mayne, D.Q., E. Polak and A.J. Heunis, (1979a). Solving nonlinear inequalities in a finite number of iterations. Publication 79/3, Department of Computing and Control, Imperial College, London.

Mayne, D.Q., E. Polak and R. Trahan, (1979b). An outer approximations algorithm for computer aided design problems. UCB/ERL M77/10, ERL, College of Engineering, University of California, Berkeley. Jour. of Optimization Theory and Applications; 28, No. 3, 331.

Mayne, D.Q. and E. Polak, (1978). A superlinearly convergent algorithm for constrained optimization problems. Publication 78/52, Department of Computing and Control, Imperial College, London.

Mifflin, R., (1977). Semismooth and semiconvex functions in constrained optimization. SIAM J. on Control and Optimization, 15, 959.

Polak, E., (1979). An implementable algorithm for the optimal design centering, tolerancing and tuning problem. Submitted for publication.

Polak, E. and D.Q. Mayne, (1976). An algorithm for optimization problems with functional inequality constraints. IEEE Trans. AC-21, No. 2, 184.

Polak E. and D.Q. Mayne, (1978a). On the finite solution of nonlinear inequalities. Report UCB/ERL M78/80, ERL, College of Engineering, University of California, Berkeley.

Polak, E. and D.Q. Mayne, (1978b). A robust second order method for optimization problems with inequality constraints. Report, UCB/ERL, ERL, College of Engineering, University of California, Berkeley.

Polak, E. and A. Sangiovanni-Vincentelli, (1979). Theoretical and computational aspects of the design centering, tolerancing and tuning problem. IEEE Trans. on Circ. and Syst., CAS-26, No. 9.

Polak, E., R. Trahan and D.Q. Mayne, (1979). Combined phase I-phase II methods of feasible directions, Mathematical Programming, 17, No. 1, 61.

Voreadis, A., (1978). Computer aided design of multivariable systems using a feasibility algorithm. MSc Thesis, Department of Computing and Control, Imperial College, London.

Zakian, V. and U. Al-Naib, (1973). Design of dynamical and control systems by a method of inequalities. Proc. IEE, 120, No. 11, 1421.

HIERARCHICAL CONTROL OF LARGE SCALE LINEAR SYSTEMS WITH AN APPLICATION TO ROBOTICS

W. A. Gruver*, J. C. Hedges* and W. E. Snyder**

**Department of Electrical Engineering and Graduate Program in Operations Research, North Carolina State University, Raleigh, NC 27650, USA*

***Department of Electrical Engineering, North Carolina State University, Raleigh, NC 27650, USA*

Abstract. The development and implementation of large scale nonlinear programming algorithms for decentralized optimal control of linear dynamic systems with prescribed initial and final states is described. A multi-level structure for open and closed loop control has been based on the Interaction Prediction Principle. The approach utilizes a hierarchical decomposition into an infimal level consisting of low order optimization problems which are analytically solvable, and a supremal level in which the interaction and coordination functions are updated by a successive approximation.

Development of this algorithm was motivated by a new approach to the real time control of a six-jointed robotic manipulator using distributed processing. In contrast to work by previous investigators, the approach presented in this research is based on a horizontal decomposition of the manipulator dynamics and the assignment of an individual processor to each joint. Each processor views its actuator as a second order linear system with parameters to be identified. Issues of implementation using an array of Texas Instruments TMS 9900 microprocessors are discussed.

Keywords. Optimal control, hierarchical optimization, distributed processing, robotic manipulator control, decentralized control.

INTRODUCTION

This paper describes a multi-level structure for decentralized optimal control and coordination of large scale linear dynamic systems with prescribed initial and final states. The technique is derived from the Interaction Prediction Principle, and is suitable for implementation using distributed processing. Development of the algorithm derived in the following sections was motivated by a problem of path control for a robotic manipulator system comprised of six mechanical linkages with electric and pneumatic actuators. A simplified model of the dynamics considers each joint as a second order linear system with time varying parameters and interactions to be identified. Each joint is constrained to follow a trajectory defined by a sequence of position set points. Such a problem is fundamental to treatments of controllability in linear dynamical systems [1,2], the results of which influenced our choice of criterion for the optimal control formulation in the next section.

The Interaction Prediction Principle has been treated extensively in the references [3-10], although primarily in the context of free end point problems of optimal control. An exception is Cohen [11] who studied its use for constrained optimal control as an infinite dimensional nonlinear programming problem, and derived conditions for convergence of the coordinator iteration. Compared to other approaches for hierarchical control, The Interaction Prediction Principle results in a relatively simple coordination step which can be carried out using a separate processor, or distributed among the infimal level processors at the expense of increased interprocessor communication.

In the first part of the paper we derive a two level algorithm for computing open loop control in a minimum energy sense. The last part extends the approach to obtain feedback control. Issues of implementation and computer architecture using distributed processing are summarized in the Appendix.

PROBLEM FORMULATION AND DECOMPOSITION

A continuous time formulation of the optimal control problem assumes that the system can be decomposed as follows:

Determine an optimal control $u_i \in L_2^{m_i}[0,T]$ such that

$$F(u_1, \ldots, u_M) = \frac{1}{2} \sum_{i=1}^{M} \int_0^T u_i'(t)R_i(t)u_i(t)\, dt \tag{1}$$

is minimized over all $u_i \epsilon L_2^{m_i} [0,T]$ subject to

$$\dot{x}_i(t) = A_i(t)x_i(t) + B_i(t)u_i(t) + C_i(t)v_i(t) \tag{2}$$

$$v_i(t) = \sum_{j=1}^{M} (D_{ij}(t)x_j(t) + E_{ij}(t)u_j(t)) \tag{3}$$

with boundary conditions

$$x_i(0) = x_{io} \tag{4}$$

$$x_i(T) = x_{i1} \tag{5}$$

for $i = 1, \ldots, M$ and $t \epsilon [0,T]$, where $x_i(t) \epsilon \mathbb{R}^{n_i}$, $v_i(t) \epsilon \mathbb{R}^{r_i}$ and the $m_i x m_i$ matrix $R_i(t)$ is symmetric and positive definite for each $t \epsilon [0,T]$.

The solution approach for the latter problem consists of characterizing an optimal control by first order necessary conditions, with an iterative procedure based on a relaxation scheme. Necessary conditions for an unconstrained minimum of F at (x_i, u_i, v_i) are as follows:

$$u_i(t) = -R_i^{-1}(t)(B_i'(t)p_i(t) - \sum_{j=1}^{M} E_{ji}'(t)q_j(t)) \tag{6}$$

$$\dot{p}_i(t) = -A_i'(t)p_i(t) + \sum_{j=1}^{M} D_{ji}'(t)q_j(t) \tag{7}$$

$$q_i(t) = -C_i'(t)p_i(t) \tag{8}$$

where the functions p_i and q_i represent Lagrange multipliers for the system constraint, Eq. (2), and the interaction constraint, Eq. (3), respectively. In addition, the constraints described by Eqs. (2-5) are also satisfied at the optimal point.

Following the Interaction Prediction Principle [3-5], the necessary conditions Eq. (3) and Eq. (8) are relaxed, and infimal problems are constructed whose solutions satisfy the remaining necessary conditions. Assuming for the moment that the functions v_i and q_i were known, we define the following infimal problem having necessary conditions for an unconstrained minimum which are identical to those for the relaxed problem:

Determine an optimal control $\hat{u}_i \epsilon L_2^{m_i}[0,T]$ such that

$$F_i(u_i) = \int_0^T (\frac{1}{2} u_i'(t)R_i(t)u_i(t) + \sigma_i'(t)x_i(t) + \gamma_i'(t)u_i(t))dt \tag{9}$$

is minimized over all $u_i \epsilon L_2^{m_i}[0,T]$ subject to

$$\dot{x}_i(t) = A_i(t)x_i(t) + B_i(t)u_i(t) + C_i(t)v_i(t) \tag{10}$$

$$x_i(0) = x_{io} \tag{11}$$

$$x_i(T) = x_{i1} \tag{12}$$

where $\sigma_i(\cdot)$ and $\gamma_i(\cdot)$ are defined by

$$\sigma_i(t) = - \sum_{j=1}^{M} D_{ji}'(t)q_j(t) \tag{13}$$

$$\gamma_i(t) = - \sum_{j=1}^{M} E_{ji}'(t)q_j(t) \tag{14}$$

for $i = 1, \ldots, M$ and $t \epsilon [0,T]$.

An advantage of this latter formulation is that the solution of the infimal problem can usually be obtained without iteration. A successive approximation subsequently determines improved values for the coordination and interaction functions q_i and v_i respectively.

SOLUTION OF THE INFIMAL PROBLEM

A solution of the infimal problem is obtained using results from the controllability of linear systems. Defining the transition matirx $\phi_i(t,s)$ of the ith subsystem, the solution of Eq. (2) and Eq. (7) can be represented as

$$x_i(t) = \phi_i(t,0)x_{io} + \int_0^t \phi_i(t,s) [B_i(s)u_i(s) + C_i(s)v_i(s)]ds \tag{15}$$

$$p_i(t) = \phi_i'(T,t)p_{i1} + \int_t^T \phi_i'(s,t)\sigma_i(s)\, ds \quad (16)$$

where p_{i1} is the unknown boundary condition for Eq. (7).

Solving Eqs. (16-17) and Eq. (6) for p_{i1} yields after a brief calculation, the following expression:

$$p_{i1} = W_i(T,0)^{-1}\eta_i(0) \quad (17)$$

where

$$W_i(T,0) = \int_0^T \phi_i(T,s)B_i(s)R_i^{-1}(s)B_i'(s)\phi_i'(T,s)ds \quad (18)$$

$$\eta_i(0) = \phi_i(T,0)x_i(0) - \int_0^T \phi_i(T,s)B_i(s)R_i^{-1}(s) B_i'(s) \int_s^T \phi_i'(\tau,s)\sigma_i(\tau)d\tau ds + \int_0^T \phi_i(T,s)B_i(s)\cdot R_i^{-1}(s)\gamma_i(s)ds + \int_0^T \phi_i(T,s)C_i(s)v_i(s)ds - x_i(T). \quad (19)$$

The existence of the inverse in Eq. (17) is guaranteed if the ith subsystem is completely controllable. If the inverse fails to exist, the infimal problem may or may not have a solution, i.e., there may be no solution or there may be infinitely many solutions for p_{i1}.

In both cases, the existence and uniqueness in a "smallest least squares" sense may be ensured by use of the Moore-Penrose pseudo inverse [12].

THE DECENTRALIZED CONTROL ALGORITHM

The solution of the infimal problem may be obtained by a two-level hierarchical algorithm as follows:

Algorithm

(1) Choose $\varepsilon_i > 0$, $\alpha\epsilon(0,1)$, $\beta\epsilon(0,1)$, v_i^o and q_i^o for $i = 1, \ldots, M$; set $k = 0$.

(2) With $v_i = v_i^k$ and $q_i = q_i^k$, solve the infimal problem using Eqs. (6) and (16-17) to obtain $x_i = x_i^k$, $u_i = u_i^k$ and $p_i = p_i^k$.

(3a) Compute v_i^{-k} and q_i^{-k} by

(3) a. Compute $\bar{v}_i^k$ and $\bar{q}_i^k$ by

$$\bar{v}_i^k(t) = \sum_{j=1}^{M} (D_{ij}(t)x_i^k(t) + E_{ij}(t)u_i^k(t))$$

$$\bar{q}_i^k(t) = -C_i'(t)p_i^k(t), \text{ for all } t\epsilon[0,T].$$

b. Evaluate

$$v_i^{k+1}(t) = \alpha v_i^k(t) + (1-\alpha)\bar{v}_i^k(t)$$

$$q_i^{k+1}(t) = \beta q_i^k(t) + (1-\beta)\bar{q}_i^k(t),\ t\varepsilon[0,T].$$

(4) If $||v_i^{k+1} - v_i^k||^2 + ||q_i^{k+1} - q_i^k||^2 \leq \varepsilon_i$ for $i = 1, \ldots, M$ stop.

(5) Increase k by one and go to step (2).

The use of the averaging parameters α and β in step (3b) increases the radius of convergence of the algorithm with respect to interactions, time horizon and initial conditions. As in other algorithms of this type, the stopping criterion in step (4) can be replaced by other suitable criteria.

DECENTRALIZED PARITAL FEEDBACK CONTROL

In many applications, a feedback structure for the decentralized control is important. Closed loop control, for example, is essential in robotics, due to imprecise knowledge of the system parameters and distrubances. This section describes a simple modification of the open loop procedure treated in preceding sections to obtain "partial feedback" control.

An explicit form for the initial feedback control can be developed by substituting Eq. (17) into Eq. (6) to obtain

$$u_i(0) = K_i(0)x_i(0) + y_i(0) \quad (20)$$

where $K_i(0)$ is an $m_i x n_i$ feedback gain matrix and $y_i(0)$ is an m_i - vector dependent on the boundary conditions $x_i(0)$ and $x_i(T)$. A feedback control can be obtained by replacing the initial time by the current time in Eq. (2), and solving a minimum energy problem over the remaining time interval [t,T] to obtain

$$u_i(t) = K_i(t)x_i(t) + y_i(t) \quad (21)$$

where

$$K_i(t) = -R_i^{-1}(t)B_i'(t)\phi_i'(T,t)W_i^{-1}(T,t)\phi_i(T,t) \quad (22)$$

$$y_i(t) = -R_i^{-1}(t)\{B_i'(t)[\phi_i(T,t)W_i^{-1}(T,t)(n_i(t) - \phi_i(T,t)x_i(t)) - \int_t^T \phi_i'(s,t)\sigma_i(s)ds] + \gamma_i(t)\} \quad (23)$$

A sufficient condition that a unique feedback control exist is that the $n_i x n_i$ matrix $W_i(T,t)$ be positive definite for each $t\epsilon[0,T]$. However, for computational purposes, the actual determination of W_i is facilitated by consideration of constant systems, since the transition matrix can be evaluated by a matrix exponential. Time varying systems can be directly treated by a discretization of the continuous problem, derivation of the corresponding discrete necessary conditions, and a finite dimensional search for the interaction and coordination variables.

CONCLUSIONS AND COMMENTS

An algorithm for decentralized minimum energy control of large scale, linear, interconnected systems has been derived based on the Interaction Principle. In analogy to Takahara [4], conditions ensuring the convergence of the successive approximation can be obtained by consideration of the underlying fixed point problem.

Numerical comparisons by the authors show that the convergence of the algorithm, when implemented for piecewise constant controls, is insensitive to the sampling interval of the discretization, however can be sensitive to the interactions and time horizon. Experiments have demonstrated that constant averaging with a parameter choice of 0.24 to 0.5 can aid the convergence of the successive approximation for the coordinator.

Partial feedback control is obtained by regarding the initial state as the current state and minimizing over the remaining time interval. The resulting feedback gain can be easily computed by evaluation of certain controllability matrices, without iteration at the infimal level. Although other criteria may be considered, the simplicity of the infimal problem is a major advantage of the minimum energy formulation. The computational overhead associated with treating deviations from intermediate points along a desired path by use of a penalty function would be prohibitive for online implementation in most manipulator systems. The iterative nature of the coordination step of the proposed algorithm is the main disadvantage in comparison to other decentralized schemes under investigation. Nevertheless, the results of related studies [20] suggest that distributed processing is feasible for implementation of optimal path control of medium size robotic systems.

Appendix

DESCRIPTION OF THE DISTRIBUTED PROCESSING SYSTEM

Distributed processing has different meanings depending on the viewpoint of the potential user. In the context of process control, a distributed processing system uses more than one computer to control subsystems of a process. These individual processors constitute a network of computers which can be used to obtain coordinated control over the entire process using a suitable algorithm.

One motivation for adopting a distributed processor architecture for a control system is modularity, both functional and physical. Functional modularity reflects the tendency of the system to remain constant over many possible system sizes. Suppose for instance we have a six-jointed manipulator, wish to add a seventh joint and have a good solution to the distributed control problem. We need only add one more controller, identical to all the others, rather than redesigning or re-programming the entire system. This functional modularity concept applies not only to controlling general purpose manipulators but also to adding a new actuator to a complex system of actuators performing some other assembly line or process control function.

Physical modularity reflects the fact that we are trading software cost for hardware cost. High volume, low cost microprocessors offer a high degree of functional and physical modularity. In contrast, a single processor for all the actuators requires a complex time slicing arrangement to insure that each actuator is serviced in sufficient time. The latter is subject to severe race conditions; for example, there may be times when it is essential that the computer monitor very closely the activity of a particular set of actuators, and cannot afford to relinquish time to maintain current "world model" data base, or to handle communications with the operator. Of course, such hazards have a low probability of occurence since computers are typically much faster than mechanical time constraints; however, the software effort to schedule the system so as to avoid such conditions can be very extensive. By using separate processors for each servo, the scheduling task can be allocated to another physical machine. This example is only one of many potential means whereby hardware and software costs may be traded. The presence of physical modularity also provides greatly improved fault detection and fault tolerance, fail soft ability, ease of maintainance and higher reliability.

If a control module fails, typically only the behavior of the particular actuator controlled by that module is affected. Such failures are less likely to be disasterous in a global sense than a corresponding failure in a single processor system. The ability to "module swap" provides for ease of maintenance and low down time. It is even possible to have processors which are currently idle to monitor the progress of other processors and provide early and rapid fault detection. The general concepts of distributed computing are discussed in Ref. [13-16].

A manipulator provides an excellent environment in which to evaluate distributed computing condepts, since a manipulator requires precise real time coordination of the processors, and the success or failure of such concepts is readily observable. Once we have made the committment to distributed control, we are naturally led to a set of guidelines which are consistent with that decision. These guidelines are as follows:

1) Servoing should be done in joint space with trajectory endpoints specifying both position and velocity of joint variables [17]. Paul has shown that this is reasonable even for tasks such as moving on a conveyor.

2) Implementation of an accurate model for the dynamics of the entire manipulator should be avoided, since such models, even with many simplifying assumptions are usually computationally impractical.

3) Each joint should be controlled optimally by a separate processor.

4) Changes in arm configuration should be treated as constants in the dynamic model of the joint (allowing simplified optimal control); however, the parameters should be rapidly and dynamically estimated and updated.

5) Each processor should be responsible for control of one and only one joint, and for reasons of modularity, maintainability, the processors should be as similar as possible.

6) Each processor should monitor the functioning of the other processors and have certain override capabilities so as to make the system failure tolerant.

The development of our distributed processing system for the control of robotic manipulator has as its basis the concept of a multi-echelon structure [3]. Each separate processor provides coordination and control of a single joint of the robot manipulator and has a means of communication with its associated processors. This scheme can be compared with the vertical decomposition pipelining proposed by Luh [18], in which the processors are assigned tasks by computational function, such as computing all the kinematic transformations, while other processors compute the controls. The communications path in our system is a shared or global memory. In addition to the memory, communication between processors is available over a relatively slow serial data link, which serves as a backup communications path to allow controlled shutdown in the case of failure of the memory.

The processing system is based on the Texas Instruments TMS 9900 which offers an advantageous mix of speed, funciton, and bandwidth. Operating with the TMS 9900's is a simple interface between the individual processor and the global memory. The arbitration for access to the global memory bus is performed by a multi-phase clock.

The arbitration scheme provides a mechanism whereby no processor speed is lost due to memory arbitration conflicts. The basic scheme makes use of very fast MOS RAM. By assigning to each processor a specific window in time, and by making those windows very short, it is possible to insure that each processor gets a memory cycle whenever it needs it. In addition, the interfaces have the ability to be expanded to more processors without limit, sacrificing performance only as the number of processors multiplied by the speed of the clock exceeds the speed of memory. The processing system consists of multiple microprocessors each of which has dedicated RAM, ROM, I/O ports and an interface to the global memory. Using the taxonomy of Anderson and Jensen [19], the system is a Direct Shared Memory (DSM) organization. As is characteristic with a DSM system, the processing system offers excellent modularity with respect to the processor. A direct dedicated path to the memory has been provided for each processor allowing utilization of the available memory bandwidth and contributing a high degree of fault tolerance.

Further details of the architecture and experimental results based on the system are presented in the reference [20].

References

[1] Kalman, R. E., Ho, Y. C., and Narendra, K. "Controllability of Linear Dynamical Systems," in Contributions to Differential Equations, Vol. 1, pp. 189-213, 1963.

[2] Padulo, L., and Aribib, M., Systems Theory, Saunders Publishing Company, Philadelphia, 1974.

[3] Mesarovic, M. D., Macko, D., and Takahara, Y., Theory of Hierarchical Multilevel Systems, Academic Press, NY, 1970.

[4] Takahara, Y., A Multilevel Structure for a Class of Dynamic Optimization Problems, Master's Thesis, Engineering Division, Case Western University, Cleveland, OH, 1965.

[5] Brosilow, C. B., Lasdon, L. S., and Pearson, J. D., "Feasible Optimization Methods for Interconnected Systems," Proceedings of the Joint Automatic Control Conference, Troy, NY 1965.

[6] Pearson, J. D., "Dynamic Decomposition Techniques," in Optimization Methods for Large Scale Systems, D. A. Wismer (ed.), McGraw-Hill Book Company, New York, pp. 121-290, June, 1977.

[7] Cohen, G., "On an Algorithm of Decentralized Optimal Control," J. Math. Anal. and Appl., Vol. 59, pp. 242-259, June, 1977.

[8] Hakkala, L., and Hivonen, J., "On Coordination Strategies for the Interaction Prediction Principle Using Gradient and Multiplier Methods," Int. J. Systems Sci., Vol. 9, No. 10, pp. 1179-1195, 1978.

[9] Grateloup, G., and Titli, A., "Two-Level Dynamic Optimization Methods," J. Optimiz. Theory Appl., Vol. 15, No. 3, 1975.

[10] Singh, M. G., Hassan, M., and Titli, A., "Multi-Level Control of Interconnected Dynamical Systems Using the Prediction Principle," IEEE Trans. Systems, Man and Cyb., Vol. SMC 5, pp. 233-239, 1976.

[11] Cohen, G., and Joalland, G., "Coordination Methods by the Prediction Principle in Large Dynamic Constrained Optimization Problems," Proceedings of the IFAC Symposium on Large Scale Systems, Udine, Italy, pp. 539-547, 1976.

[12] Dorny, C. N., A Vector Space Approach to Models and Optimization, John Wiley and Sons, New York, NY 1975.

[13] Enslow, P. H., "What is a Distributed Processing System?," Computer, V9, No. 1, January, 1978.

[14] Enslow, P. H., Ed., Multiprocessors & Parallel Processing, Wiley, 1974.

[15] Julissen and Mowle, "Multiple Microprocessors with Common Main and Control Memories," IEEE Trans. on Computers, V. C-26, No. 11, November, 1973.

[16] Jensen, E. D., "Distributed Processing in a Real-Time Environment," Infotec State of the Art Report, November, 1975.

[17] Paul, R., "Cartesian Coordinate Control of Robots in Joint Coordinants," Third CISM-IFTOMM International Symposium on Theory and Practice of Robots and Manipulators, Udine, Italy, 1978.

[18] Luh, J. Y. S., Voli, R. P., and Walker, M. W., "Mechanical Arm with Microcomputer as Controller," Int. Conf. on Cybernetics and Society, Washington, DC, 1976.

[19] Anderson and Jensen, "Computer Interconnection Structures, Taxonomy, Characteristics, and Examples," ACM Computing Surveys, Vol. 7, No. 4, December, 1975.

[20] Snyder, W. E., Evans, P. F., and Gruver, W. A., "A Shared Memory Architecture for a Distributed Microcomputer Control System," Computer Studies Tech. Report, TR-78-17, North Carolina State University, Raleigh, NC, February, 1979.

OPTIMAL AND SUBOPTIMAL CONTROL OF OSCILLATING DYNAMICAL SYSTEMS

F. L. Chernousko, V. M. Mamaliga and B. N. Sokolov

Institute for Problems of Mechanics, USSR Academy of Science, Moscow, USSR

Abstract. Problems of control for dynamical systems with oscillating elements arise in many fields of technology, for instance, control of elastic systems, control of cranes carrying pendulous loads, control of robots and manipulators. This paper is devoted to investigation and solution of some typical problems of optimal control for oscillating systems.

Keywords. Optimal Control, Oscillating Systems

INTRODUCTION

We consider optimal motions of oscillating systems with controlled equilibrium position. The time-optimal control problems for such systems were considered earlier in the papers (Anselmino, Liebling, 1967; Beeston, 1967; Flugge-Lotz, Mih Yin, 1961; Flugge-Lotz, Titus, 1962; Hippe, 1970; Mikhailov, Novoseltseva, 1964, 1965; Moros, 1969; Troitskii, 1976). These papers are concerned with questions of optimal (minimum-time) transient processes in linear systems of the 3d and 4th order.

In this paper we study optimal motions of oscillating systems which consist of a pendulum attached to a rigid body. The results obtained below can be applied also for the control of other oscillating systems such as systems containing elastic elements or fluid in a tank.

Different problems of minimal time transfer of such systems between two prescribed states are solved under various boundary conditions and constraints imposed on the control functions and phase coordinates. The conditions imply that there are no oscillations in final states of the system.

Optimal control laws are obtained analytically in the form of programme (open-loop control). In the cases when velocity or acceleration are restricted the optimal regimes are of bang-bang type and consist of several intervals of motion with constant velocity (or acceleration). The number of switching points depends on the prescribed displacement. Besides the exact optimal solutions some suboptimal regimes were found which are simpler and more convenient for practical applications. Problems of feedback control for optimal transfer are discussed also. The optimal starting and braking of oscillating systems are solved for different cases. Control problems for a pendulum with varying length are considered also. In the last case the system has two control functions.

The results given below in this paper were obtained in the Institute for Problems of Mechanics, USSR Academy of Sciences; part of them is presented in the papers (Banichuk, Chernousko, 1975; Chernousko, 1975, 1977; Sokolov, 1977; Sokolov, Chernousko, 1976, 1977; Zaremba, Sokolov, 1978, 1979).

DESCRIPTION OF SYSTEMS

We considered a controlled oscillating system consisting of a pendulum P attached to a rigid body G. The masses of the bodies P, G are m and M respectively. The body G can move with velocity V along the horizontal axis X while the pendulum P swings about the horizontal axis O with the body G (Fig. 1).

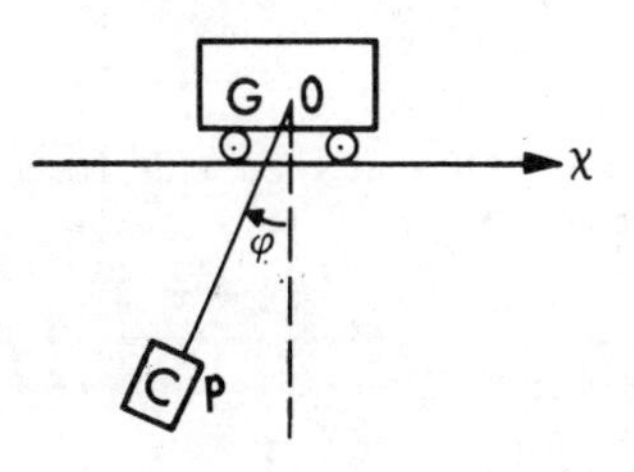

Fig. 1

The equations of motion of the system for small plane oscillations are

$$(M + m)\ddot{x} - mL\ddot{\gamma} = F$$
$$I\ddot{\gamma} + mgL\gamma = mL\ddot{x} \qquad (1)$$

Here x is a coordinate of the centre of mass of the body G, γ is an angle between OC and the vertical, L is a distance OC between the axis O and the centre of mass C of the pendulum, F is a force control applied to the body G, I is a moment of inertia of the pendulum about the axis 0, g is the gravity acceleration. The motion of the body G is controlled by the force F which is bounded by the condition

$$|F(t)| \leq F_o \tag{2}$$

where F_o is a constant. We introduce the following dimensionless variables x', t', ϕ', u according to the formulae:

$$x'=T_o^{-2}F_o^{-1}((M+m)x-mL\ \phi),\quad t'=T_o^{-1}t$$

$$\phi'=(M+m)gF_o^{-1}\phi,\quad u = F\ F_o^{-1}$$

$$T_o=(I/(mgL) - mL(M=m)^{-1}g^{-1})^{1/2} \tag{3}$$

Eqs. (1) after the transformation (3) assume the form:

$$\dot{x}=v,\ \dot{v}=u,\ \dot{\phi}=\omega$$
$$\dot{\omega} = -\phi + u,\ |u| \leq 1 \tag{4}$$

where symbols (') are omitted for simplicity.

In the case when the central function is a velocity v of the body G, then equations of motion are given by

$$I\dot{\phi} + mgL\phi = mLw$$
$$\dot{x} = v,\quad \dot{v} = w \tag{5}$$

Here w is the horizontal acceleration of the body G. Its velocity is bounded by the condition

$$-v_o\gamma \leq v \leq v_o \tag{6}$$

where v_o, γ are given constants. When $\gamma= 0$ we have the following asymmetrical constraint $0 \leq v \leq v_o$ which does not allow a reverse motion of the body G. For system (5) and (6) we introduce dimensionless variables

t'', x'', v'', w'', ϕ" by the formulae

$$t = T_1t'',\ x = v_oT_1x'',\ v = v_ov''$$
$$w = v_oT_1^{-1}w'',\ \phi = v_oT_1^{-1}g^{-1}\phi'' \tag{7}$$
$$t_1 = I^{1/2}/(mgL)$$

and transpose (5) - (6) to the form (symbols ('') are omitted)

$$\dot{\phi}= \omega,\ \dot{\omega}= -\phi + w,\quad x = v$$
$$\dot{v}= w,\ -\gamma \leq v \leq 1 \tag{8}$$

Now, using variable $\Psi = v - \omega$, we can transform Eqs. (8) to the following system:

$$\dot{\Psi} = \gamma,\ \dot{\gamma}= -\Psi + v,\quad \dot{x} = v$$
$$\dot{v} = w,\ -\gamma \leq v \leq 1 \tag{9}$$

Here Eq. $\dot{v} = w$ may be omitted. Thus we have

$$\Psi = \gamma, \dot{\gamma} = -\Psi + v,\ \dot{x}=v$$
$$-\gamma \leq v \leq 1 \tag{10}$$

where v is a control function.

Eqs. (4) and (10) were derived above for particular oscillating system with pendulum, but they are valid also for other controlled systems of various physical nature, such as systems containing elastic elements or fluid in a tank. Some examples are presented at Fig.2 and Fig. 3. The system of Fig. 2 consists of two mass which move along the horizontal axis Ox.

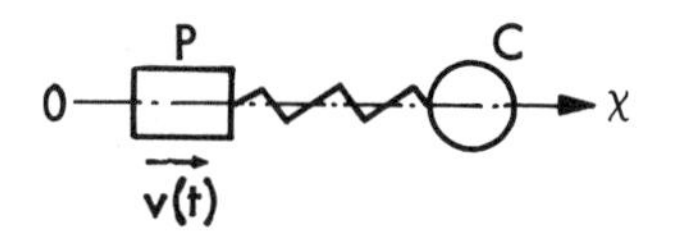

Fig. 2

Velocity v of the point P is a control function and the mass C is connected with the point P by a linear spring. If we denote by γ the displacement of the mass 0 with respect to its current equilibrium position, the equations of motion for this system are given by (8). The second example is shown at Fig. 3.

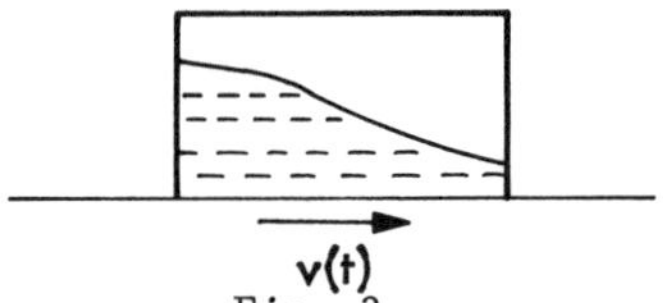

Fig. 3

Here we have a tank containing ideal incompressible fluid moving with velocity v along the horizontal axis. If we consider only principal mode of small oscillations of fluid, then equations of motion are also given by (8). Another example is the case of torsional controlled oscillations. We consider below some time-optimal control problems for systems of the types (4), (8), (10).

OPTIMAL TRANSFER WITH BOUNDED VELOCITY

We begin with the problem of optimal time motion of the system (8) between two given states of rest. The boundary conditions are

$$\gamma(o) = \omega(o) = x(0) = v(0) = 0$$
$$\gamma(T) = \omega(T) = 0,\ x(T) = a,\ v(T) = 0 \tag{11}$$

Here a is a given dimensionless displacement of the system. The velocity v is subjected to constraint $-\gamma \leq v \leq 1$. Boundary conditions (11) in terms of γ, Ψ, x for the system (10) have the following form

$$\Psi(0) = \gamma(0) = x(0) = 0$$

$$\Psi(T) = \gamma(T) = 0,\ x(T) = a$$

It was found (Chernousko, 1975; Sokolov, Chernousko, 1976; Mamaliga, 1978) that the optimal velocity law of the oscillator for this time-optimal problem is determined by a bang-bang type formula:

$$v(t) = \{(1-\gamma)+(-1)^{t+i}(1+\gamma)\}/2$$
$$\sum_{j=1}^{i-1} t_j < t < \sum_{j=1}^{i} t_j, \tag{13}$$
$$i = 1,2,\ldots,n$$

In order to obtain the number of intervals n and their durations t_i in the formula (13), it is necessary to represent a in the form $a = 2\pi k + b$, where $k \geq 9$ is a positive integer and $0 \leq b < 2\pi$. If $b = 0$ then we have a simple solution $n = 1$, $v = 1$, $T = a$. In this case the optimal motion has constant velocity $v = 1$. In the general case $b \neq 0$ we have $T > a$, $b > 0$, and here $n = 2k + 3$. The values of t_i are expressed by means of the unique root τ of the Eq.

$$b=\tau -2(\gamma+1)(k+1)\arcsin \frac{\sin(\tau/2)}{(k+1)(\gamma +1)} \quad (14)$$

When the value of τ is determined through Eq. (14), the values of t_i and the total optimal time T are given by

$$T = 2\pi k + \tau,\ t_1 = t_n = \tau/2-\alpha_k$$
$$t_3=t_5=\ldots=t_{n-2}=2(\pi-\alpha_k),\ n-2k+3 \quad (15)$$
$$t_2=t_4=\ldots=t_{n-1}=2\alpha_k$$
$$\alpha_k=\arcsin(k+1)^{-1}(\gamma +1)^{-1}\sin(\tau /2))$$

The solution of the considered optimal control problem is completely determined for all $a \geq 0$ and $\gamma \geq 0$. The suboptimal controls are built also which have a small number of switching points. The simplest suboptimal solution is the control with three intervals weere the velocity is constant. This suboptimal control is useful for practical applications. Its total time exceeds the minimal time by less then 1.2% for any $\gamma \geq 1$ and arbitrary a, see ref. (Chernousko, 1975; Sokolov, Chernousko, 1976).

OPTIMAL TRANSFER FEEDBACK CONTROL WITH BOUNDED VELOCITY

In the previous section we considered optimal time transfer of the oscillating system (8) or (10) between two given states of rest. Here the solution for optimal transfer is given with arbitrary initial position of the system. The initial conditions at t = 0 for the system (8) are

$\gamma(0) = \gamma_0$, $\omega(0) = \omega_0$, $x(0)= v(0) = 0$. At the final moment $t = T$ we have conditions $\gamma(T) = \omega(T) = 0$, $x(T) - a$, $v(T) = 0$.

In the paper (Mamaliga, 1979) the surface of switching which gives the optimal feedback control is obtained in the parametrical form. This surface was calculated numerically by A.T.Zaremba and is shown in Fig. 4

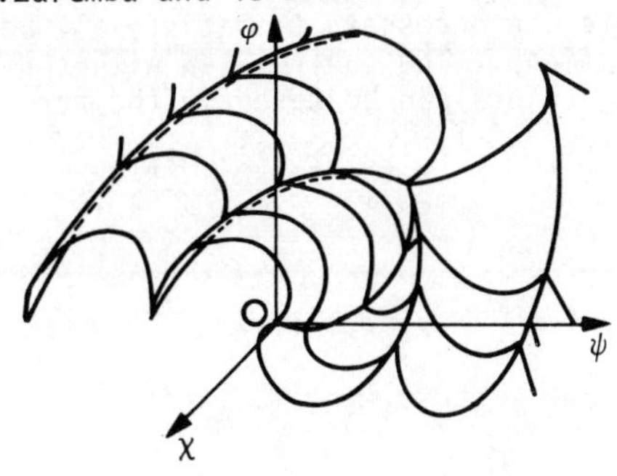

We present here some analytical results for important particular cases and some suboptimal regimes. Let the system (8) at the time $t = 0$ be at the position $\gamma= \gamma_0$, $\omega =\omega_0$, $x = v = 0$. It is required to transfer the system to the position $\gamma = \omega = v = 0$, $x = a$. It can be shown (Mamaliga (1979)) that the maximal and minimal a for which this transfer is possible during the time $T = T_s = 2\pi s$, $s = 1,2,\ldots$ are given by

$$\max(x(2\pi s))=a_s^* =2\pi s-2(1+\gamma) sc,\ s=1,2,\ldots$$

$$\min(x(2\pi s))=a_s^{**}=-2\pi\gamma s+2(1+\gamma)sc$$

$$c=\arcsin((\gamma_0^2+\omega_0^2)^{1/2}(2s(1+\gamma))^{-1}) \quad (16)$$

Functions $a_s^*(z)$, $a_s^{**}(z)$ for $\gamma = 1$ and $\gamma = 0$ are represented at Fig. 5 and Fig. 6 respectively.

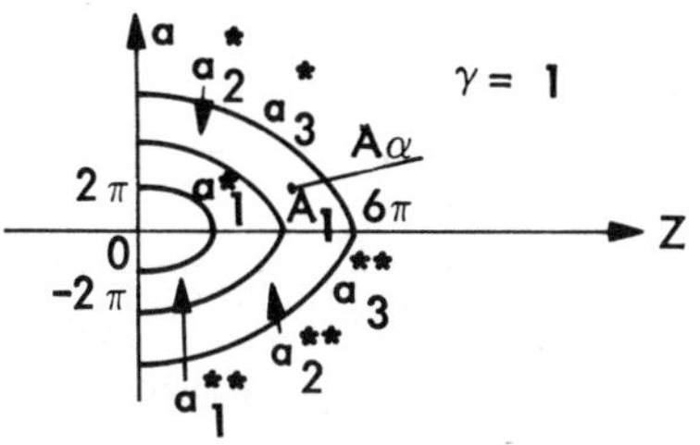

Fig. 5

Here z is equal to $\pi(\gamma_0^2 + \omega_0^2)^{1/2}/(1+\gamma)$.

The Eqs. (16) determine the boundary of the accessible set for the system (8) at the time moment $t = T_s$. These equalities can serve as useful guide to the evaluation of optimal time transfer.

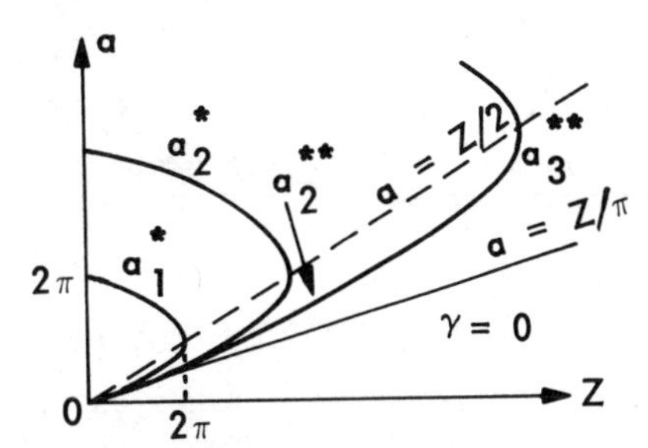

Fig. 6

Let c is given by Eq. (16) and $d = \arcsin(|\omega_0|/\sqrt{\gamma_0^2 + \omega_0^2})$. We can write now optimal control for such values γ_0, ω_0, a that satisfy formulae (16), i.e. $a = a_s^*$. The control v_0, v_1 and parameters k = 0, 1 given below correspond to regimes of optimal control with $a = a^*$ and $a = a^{**}$ respectively. These controls are determined by the following expressions:

a) If $(-1)^k \gamma_0 \geq 0$, $(-1)^k \omega_0 \geq 0$, $2(-1)^k\omega_0(1+\gamma)s \geq \gamma_0^2 + \omega_0^2$ then control v_0 is given by

$$v(t)=(1-\gamma+(-1)^{i+1}(1+\gamma))/2 \quad (17)$$

$\sum_{j=1}^{i-1} t_j < t < \sum_{j=1}^{i} t_j$, $i=1,2...,2s+1$

and v_1 is expressed by

$$v(t)=(1-\gamma+(-1)^i(1+\gamma))/2 \quad (18)$$

$\sum_{j=1}^{i-1} t_j < t < \sum_{j=1}^{i} t_j$,

$i=1,2,...,2s+1$.

Here durations of time intervals are given by formulae $t_1=2\pi-(c+d)$, $t_2=2c$, $t_{2s+1}=d-c$ and $t_2=t_4=...=t_{2s}$; $t_3=2\pi-t_2=t_5=...$

$$...=t_{2s-1} \quad (19)$$

b) If $(-1)^k \gamma_0 \geq 0$, $(-1)^k \omega_0 \geq 0$, $2(-1)^k \omega_0(1+\gamma)s \leq \gamma_0^2+\omega_0^2$, then v_0 has the form of (8) and v_1 is expressed by (17), where durations of intervals are determined according to formulae (19) and $t_1=c-d$, $t_2=2(\pi-c)$, $t_{2s+1}=c+d$.

c) If $(-1)^k \gamma_0 \leq 0$, $(-1)^k \omega_0 \geq 0$, then v_0 has the form of (17) and v_1 is expressed by (18) where durations of intervals are determined according to formulae (19) and $t_1=\pi-c+d$, $t_2=2c$, $t_{2s=1}=\pi-(c+d)$.

d) If $(-1)^k \gamma_0 \leq 0$, $(-1)^k \omega_0 \leq 0$, then v_0 has the form of (17) and v_1 is expressed by (18) where durations of intervals are obtained according to formulae (19) and $t_1=\pi-c+d$, $t_2=2_c$, $t_{2s+1}=\pi-(c+d)$.

d) If $(-1)^k \omega_0 \leq 0$, $(-1)^k \omega_0 \leq 0$, then v_0 has the form of (17) and v_1 is expressed by (18) where durations of intervals are obtained according to formulae (19) and $t_1=\pi-c+d$, $t_2=2_c$, $t_{2s+1}=\pi+d-c$.

e) If $(-1)^k \gamma_0 \geq 0$, $(-1)^k \omega_0 \leq 0$, $2(-1)^{k+1} \omega_0(1+\gamma)s \geq \gamma_0^2+\omega_0^2$, then v_0 has the form of (17) and v_1 is expressed by (18) where durations of intervals are given by formulae (19) and $t_1=d-c$, $t_2=2c$, $t_{2s+1}=2\pi-(c+d)$.

f) If $(-1)^k \gamma_0 \geq 0$, $(-1)^k \omega_0 \leq 0$, $2(-1)^{k+1} \omega_0(1+\gamma)s \leq \gamma_0^2+\omega_0^2$, then v_0 has the form (18) and v_1 is expressed as (17), where durations of intervals are given by formulae (19) and $t_1=c+d$, $t_2=2(\pi-c)$, $t_{2s+1}=c-d$.

We shall now present a simple suboptimal feedback control law which has the following property. Its dimensionless time T_* does not exceed minimal time T more than by 2π. We introduce the following auxiliary constraint

$$-\alpha v_0\delta \leq v \leq \alpha v_0 \quad (20)$$

where $\alpha < 1$. In the case $\alpha=1$ the condition (20) is equivalent to (6). In general case $\alpha < 1$ the procedure similar to (7) for the system (5) with constraints (20) leads to the following dimensionless parameters

$$\gamma_0(\alpha)=\alpha^{-1}\gamma_0, \quad \omega_0(\alpha)=\alpha^{-1}\omega_0, \quad a(\alpha)=a\alpha^{-1} \quad (21)$$

which are similar to γ_0, ω_0, a. The functions $z(\alpha)=\pi(\gamma_0^2(\alpha)+\omega_0^2(\alpha))^{1/2}(1+\gamma)^{-1}$ and $a(\alpha)$ from Eqs. (21) for different $\alpha \leq 1$ determine a line in the plane (z,a). The points of this line depending on α we shall denote by A_α. By means of the family of curves $a=a^*(z)$, $a=a^{**}(x)$, see Fig. 5, we determine the point A with minimal α which belongs one of these curves. Thus we obtain the integer s and the value α that describe our suboptimal regime. It is calculated by means of the same formulae as the optimal regime for $T=T_s$ given above, only instead of γ_0, ω_0, a we must use the quantities (21). It can be shown that such suboptimal regime is unique and its time $T_*=T_s$ exceeds minimal time T by less than 2π. For particular cases when A_α for $\alpha=1$ belongs to the curves $a=a^*(z)$ this suboptimal regime coincides with optimal time control.

OPTIMAL TRANSFER WITH BOUNDED FORCE

We consider now time-optimal transfer of the system (4) with bounded force $|u| \leq 1$. The initial and final states are given by $x(0)=v(0)=\gamma(0)=\omega(0)=0$ and $v(T)=\gamma(T)=\omega(T)=0$, $x(T)=a$. It is obvious that the optimal law is of bang=bang type. In the paper Banichuk, Chernousko (1975) the optimal control laws with one internal switching point at t=T/2 were found which are valid only for special values of $a=4\pi^2k^2$, k=1,2,... . In general case, at least three internal switching points are necessary to satisfy all boundary conditions. The control law with three switching points can be taken in the form

$$u=(-1)^{i+1}, \quad \sum_{j=1}^{i-1} t_j < t < \sum_{j=1}^{i} t_j, \quad i=1,2,3,4 \quad (22)$$

$$t_1=t_4=T/2-\chi, \quad t_2=t_3=\chi$$

The values of T, χ are given by equations

$$T=2(a+2\chi^2)^{1/2} ,$$
$$\chi = \text{arc cos } \{\cos^2(T/4)\} \quad (22')$$

which have the unique solution for all $a \geq 0$.

It can be shown that the control law (22), (22)satisfies the boundary conditions. The adjoint variables corresponding this solution were determined, and it was shown that all conditions of Pontryagin's maximum principle are satisfied here. Therefore, the law (22), (22)' is time-optimal control law. In the particular case $a=4\pi^2 k^2, k=1,2,...$ - this law gives $\chi=0$, $T=4\pi k$ and $t_2=t_3=0$. In this case it coincides with the control law with one switching point mentioned above.

OPTIMAL START OF OSCILLATING SYSTEM

Constructed in the preceding item optimal controls provided the displacement of oscillating systems at given distance a with stopping oscillation at the end of motion. These laws depended on distance a and considered system oscillated during the displacement. Suggested here regimes of starting (braking) take the system from the position of rest to the state of translational motion (and back). These regimes are particularly convenient from practical point of view because its employment gives a possibility to divide the process of displacement into three stages: starting - the motion without oscillations - braking. The control of the starting and braking stages is determined only by the parameters of the system and does not depend on distance a . The motion without oscillation on the second (middle) stage gives a possibility to combine this kind of motion with another, for example with changing the pendulum length. It makes it possible to transfer a pendulum in a vertical plane at the given distance and height with stopping oscillations.

It is supposed that the velocity and acceleration of the equilibrium point of the system are bounded by prescribed constants α , β , b.

$$1)\ -\alpha \leq v \leq \beta \ , \quad \alpha \geq 0, \quad \beta > 0;$$
$$2)\ - b \leq w \leq 1, \quad b \geq 0 \quad (23)$$

In particular one of these restrictions can be omitted.

The motion begins from the state of rest and finishes at moment T . At this moment the velocity of the equilibrium point must be equal to constant c and oscillations must be damped. The equations and boundary conditions are the following:

$$\dot{\gamma} = \omega \ , \quad \dot{\omega} = -\gamma + w, \quad \dot{v} = w \quad (24)$$

$$\gamma(0)=\omega(0)=v(0)=\gamma(T)=\omega(T)=0, v(T)=c \quad (24)$$

The coordinate x is supposed free and the corresponding equation and boundary value in the relationship (24) are omitted.

Let us consider the following problems of pendulum time-optimal control during the time of starting.

Problem 1 (the feedback problem). By choosing control w(t) it is necessary within minimal time T to transfer the system (24) from arbitrary state $\gamma(0)$, $\omega(0)$ to final state (24) with fixed c. The velocity v(t) is bounded by the first restriction (23).

Problem 2. By choosing control w(t) it is necessary to transfer the system (24) within minimal time T from the initial state of rest to the final state (24). The acceleration w(t) is bounded by second restriction (23).

Problem 3. This problem differs from problem 2 in the following: both the velocity and acceleration are bounded by the restrictions (23).

The problem of braking is formulated similarly with obvious transposition of initial and terminal condition in (24). The problems of time-optimal braking are reduced to above considered problems of starting by the way of replacement of phase variables and inverting the time.

THE SOLUTION OF PROBLEM 1

After the introduction of variable ψ by formula $\psi = v - \dot{\gamma}$ the equations of motion and boundary conditions (24) are reduced to the following equations and boundary conditions

$$\dot{\psi} = \gamma \ , \quad \dot{\gamma} \quad - \psi + v \ ,$$
$$\gamma(0)=\psi(0)=\gamma(T)=0 \ , \quad \psi(T) = c \quad (25)$$

Here the velocity v(t) is a control function which is assumed to be restricted.

The feedback control $v(\psi, \gamma)$ transferring the system (25) from arbitrary initial state (ψ^0, γ^0) to the final one (c,0) (25) is built similarly to the feedback control from example 2 (Pontryagin and others). The curve of control switching and phase trajectories on plane ψ , γ are represented on Fig. 7.

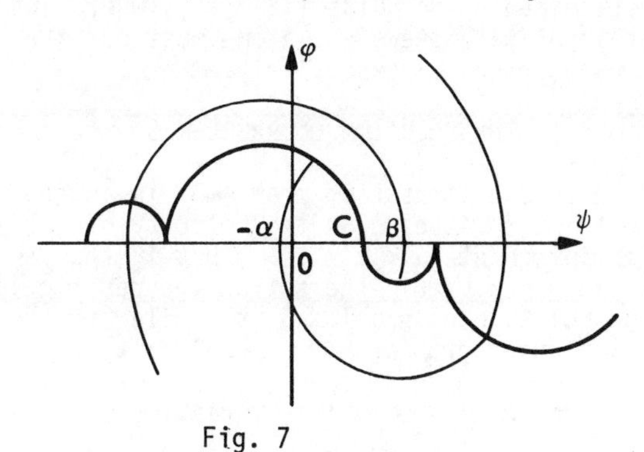

Fig. 7

From the represented feedback control it follows that the programmed control which transfers the system (25) from the state of rest to the state of motion without oscillation has two intervals of control constancy. Let the duration of the first interval be t_1 and the second one be t_2.
We have

$$v(t) = \beta \quad \text{if} \quad t \in (0, t_1);$$

$$v(t) = -\alpha \quad \text{if } t \in (t_1, t_1+t_2).$$

The case $c=\beta=1$ is being considered. In this case the mechanical system is accelerated until its velocity is equal to maximal permissible value 1. Let the parameter $\alpha=1$ or $\alpha=0$. The case $\alpha=1$ corresponds to symmetrical restriction imposed on the velocity $|v| \leq 1$, the case $\alpha=0$ corresponds to the absence of motion in opposite direction: $0 \leq v \leq 1$.

In case of $|v| \leq 1$ the duration of the intervals of the velocity constance are equal to

$$t_1=\arccos 1/4 \approx 1.3181,$$

$$t_2=\arccos 7/8 \approx 0.5054 \qquad (26)$$

$$T=\arccos(-1/4) \approx 1.8235$$

In case of $0 \leq v \leq 1$ the durations of the intervals t_1, t_2 are determined by formulas

$$t_1=t_2= \pi/3, \quad T=2\pi/3 \qquad (27)$$

THE SOLUTION OF PROBLEM 2

The problem of optimal displacement of the oscillating system at given distance a with stopping oscillations has been solved above. Let us compare the equation (10) with (24), boundary condition (12) with (24) and restriction (8) with second one (23). The relationships (8), (10), (12) are reduced to (23), (24) by the following transformation of variables

$$\psi \to \gamma, \quad \gamma \to \omega, \quad v \to w,$$

$$\dot{x} \to v, \quad \phi \to b, \quad a \to c$$

That is why the optimal control w(t) solving problems 2 is bang-bang function with the interval of constancy t_i, which durations are determined by formulas (13)-(15), where instead of parameters ϕ, a one must put the parameters b, c respectively.

THE SOLUTION OF PROBLEM 3

Problem 3 differs from problem 2 in phase restriction imposed on the velocity of equilibrium point $-\alpha \leq v \leq \beta$. Thus if the velocity lies within the set of the restriction the solution of problem 3 is similar to the solution of problem 2.

Let y be a minimum of two variables $y = \min(\tau_*, \tau^*)$, where

$$\tau_* =\{2(1+b)\beta - c\} b^{-1},$$

$$\tau^* =\{2(1+b)\alpha + (1+2b)c\} b^{-1}$$

If $y > 2\pi$ or $y < 2\pi$ and the following condition takes place

$$c \leq y-2(b+1)\arcsin (b+1)^{-1}\sin(y/2) \qquad (28)$$

then the optimal velocity v(t) lies inside the set $-\alpha \leq v(t) \leq \beta$ and optimal control w(t) is determined by formulas (13)-(15), where it is necessary to replace v, ϕ, a by w, b, c respectively.

In case when $y < 2\pi$ and inequality (28) does not take place, the optimal velocity of the equilibrium point can lie on the bound of the restriction. It has been settled that the number of time intervals, when the velocity lies on the bound, does not exceed three.

The complete analysis of admitted in this case control functions was made by Mamaliga (1978) and Sokolov (1977).

Let us consider three following important particular cases of velocity and acceleration restrictions.

A. Let the modulus of acceleration be bounded and the velocity be not less then zero $-1 \leq w \leq 1$, $0 \leq v \leq \beta$ ($\alpha=0$, b=1), $\beta \geq c$. The following is established.

a) If the parameters β, c are such that the inequalitues

$$\beta \geq c \geq 2\arcsin\{2\sin^2(\beta/2)\},$$

$$c < \pi/3$$

take place then the optimal control is (see Fig 8)

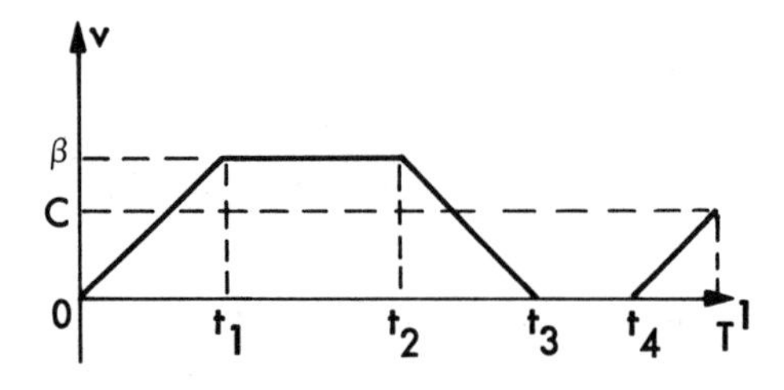

Fig. 8

$$w=1, \ t \in (0,t_1) \cup (t_4,T);$$

$$w=0, \ t \in (t_1,t_2) \cup (t_3,t_4);$$

$$w=-1, \ t \in (t_2,t_3):$$

$$t_1=\beta, \quad t_2=2\arcsin\left[\frac{\sin(c/2)}{2\sin(\beta/2)}\right]$$

$t_3=t_2+\beta$, $t_4=1/2(\pi+\beta-c+t_2)$,

$T=1/2(\pi+\beta+c+t_2)<\pi$

b) If the parameters , c are such that the inequalities

$\beta \geq c$, $c<\pi/3$,

$c \leq 2$ are sin $\{2\sin^2(\beta/2)\}$

take place then the optimal control is (see Fig. 9)

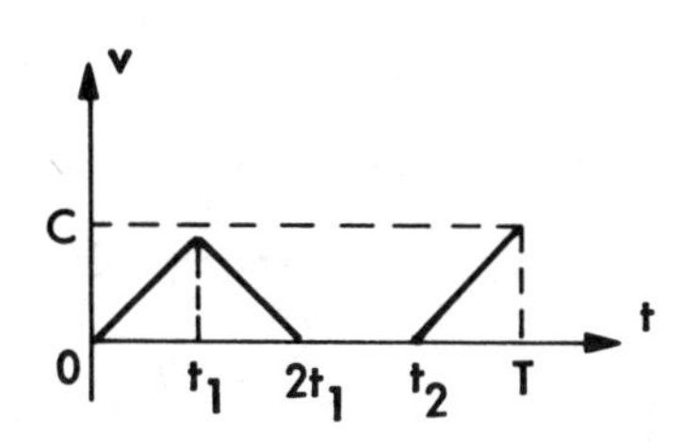

Fig. 9

$w=1$, $t \in (0,t_1) \quad U(t_2,T)$;

$w=-1$, $t \in (t_1, 2t_1)$;

$w=0$, $t \in (2t_1,t_2)$;

$t_1=2\ \text{arc sin}\ \{1/2\sin(c/2)\}^{1/2}$,

$t_2=1/2(\pi-c)+t_1$

$T=1/2(\pi+c)+t_1<\pi$

c) If $\beta \geq c \geq \pi/3$ then the solutions of problems 2 and 3 are similar.

B. Let the acceleration be bounded by an arbitrary restriction and the velocity be not less then zero

$-b \leq w \leq 1$, $- \leq v \leq \beta$ ($\alpha=0$),

$v(T)=\beta = c$

If the following inequalities

$\beta+1/2\ \beta b^{-1}<\pi$; $\beta b^{-1}<\pi$

$$2\sin\beta/2\cos\frac{\beta(1+b)}{2b} > b\sin\frac{\beta}{2b} \qquad (29)$$

take place then the optimal control is (see Fig. 10)

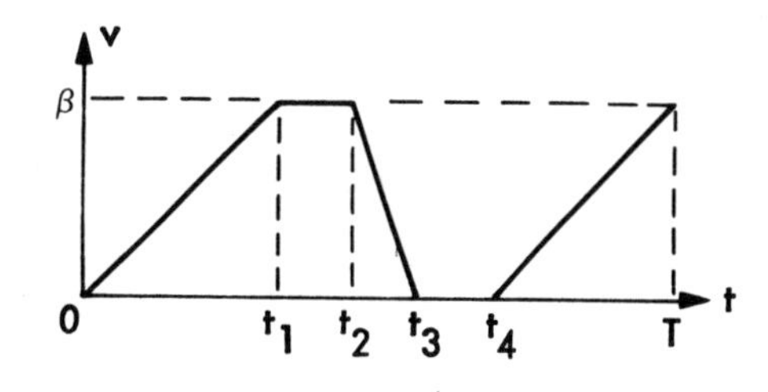

Fig. 10

$$\begin{aligned}
&w=1, && t \in (0,t_1) \quad (t_4,T);\\
&w=0, && t \in (t_1,t_2) \quad (t_3,t_4);\\
&w=-b, && t \in (t_2,t_3); \qquad\qquad (30)\\
&t_1=\beta, && t_2=T/2-\beta/2b,\ t_3=T/2+\beta/2b,\\
& && t_4=T-\beta
\end{aligned}$$

$$T=\beta+2\ \text{arc cos}\ \frac{b}{2\sin(\beta/2)}\sin\frac{\beta}{2b}$$

In limit case of $\beta\to 0$ the regime of control (30) turns into regime (27), where the velocity is assumed to be changed in given limits instantaneously.

If one or more of the inequalities (29) are violated then the optimal velocity lies inside the restrictions $0 \leq v \leq \beta$ and the solution of problems 2 and 3 are similar.

C. Let the modulus of acceleration and velocity be bounded

$-1 \leq w \leq 1$, $-\beta \leq v \leq \beta$

$(\alpha=\beta$, $b=1)$, $v(T)=c=\beta$

a) If the inequalities

$\beta_1 < \beta \leq \pi/3$, $\beta_1 \simeq 0.2548$

take place then the optimal velocity has one time interval when the value of the velocity lies on the upper bound: $v=\beta$ (see Fig. 11)

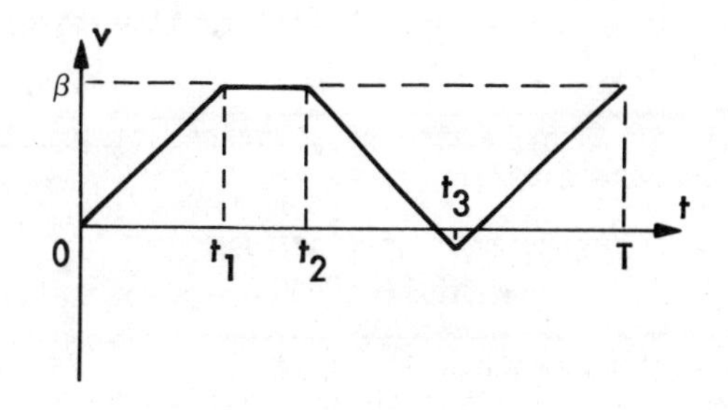

Fig. 11

Corresponding control is

$$w=1, \quad t \in (0,t_1) \quad \cup(t_3 T);$$

$$w=0, \quad t \in (t_1,t_2) \qquad (30)$$

$$w=-1, \quad t \in (t_2,t_3);$$

$$t_1=\beta, \quad t_2=\beta/2+\arcsin[1-\sin(\beta/2)],$$

$$t_3= (\pi-\beta)/2,$$

$$T=\pi+\beta/2- \arcsin \quad 1-\sin(\beta/2)$$

b) If the inequality

$$\beta \le \beta_1 \qquad (\beta_1 \simeq 0.2548)$$

takes place then the optimal velocity has the time-interval when the value of the velocity lies on the lower bound as well: $v=-\beta$(see Fig.12)

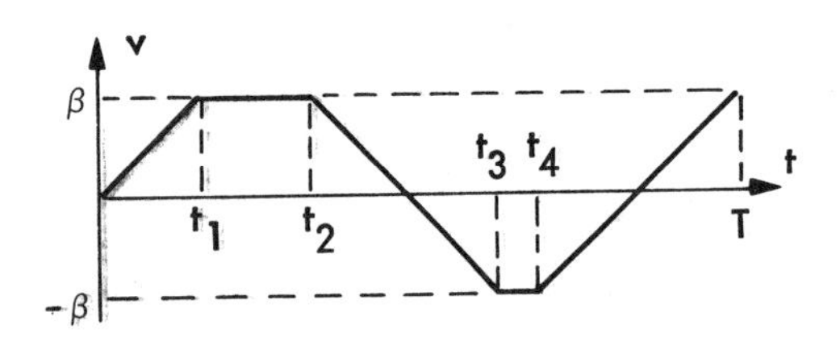

Fig. 12

$$w=1, \qquad t \in (0,t_1) \cup (t_4 T);$$

$$w=0, \qquad t \in (t_1,t_2) \cup (t_3,t_4);$$

$$w=-1, \qquad t \in (t_2,t_3); \qquad (31)$$

$$t_1=\beta, \quad t_2=\beta + x, \quad t_3=3\beta + x,$$

$$t_4 = \beta + x + 2 \arcsin y$$

$$x = \arccos y - 3/2\beta, \quad y = \{4\cos(\beta/2)\}^{-1},$$

$$T = (\pi + 3\beta)/2 = \arcsin y$$

The duration of motion is minimal if

$\beta \to 0$ and corresponding T is

$$T = \arccos(-1/4)$$

If $\beta \to 0$ then the control (31) turns into control (26).

c) If $\beta \ge \pi/3$ then the solutions of problems 2 and 3 are similar.

THE OPTIMAL DISPLACEMENT IN A VERTICAL PLANE

The systems which will be considered in this item differ from the above considered ones in an additional condition: the length of a pendulum is controlled (can be varied). The equations of motion in sizeless variables (Mamaliga, Chernousko, 1977) are

$$l\ddot{\phi} + 2\dot{l}\dot{\phi} + \phi = \ddot{x}, \quad \dot{x}=v, \quad \dot{l}=u \qquad (32)$$

Here l, u, ϕ, x, y are sizeless variables corresponding to the following dimensional variables: l - the length of a pendulum, u - the velocity of length changing, ϕ - the angle of pendulum deviation from the vertical, x - the coordinate of pendulum suspension point, v - its velocity. Controls u and v are bounded

$$0 \le u \le d, \quad 0 \le v \le 1, \quad d - \text{const}.$$

The problem of time-optimal displacement of pendulum from the state of rest with $l(0)=b\le 1$ at given distance a with damping oscillations and pendulum length $l(T)-1$ is being considerred. The simple suboptimal regimes of transferring providing the damping oscillations for any parameters a, b, d are built on the basis of above built optimal motions with constant length. Let us introduce the following denotations of the regimes of motion: A - descent (or lifting) of the pendulum without horizontal motion; B - optimal displacement at given distance a with fixed length l; C - optimal starting to maximal velocity with fixed length l_1; D - optimal braking with fixed length l_2; E - horizontal motion with maximal velocity without oscillations. Regimes ABA, ACEDA solve the problem of displacement with stopping oscillations. Regime ABA contains one free parameter, which is length l. Regime ACEDA has two free parameters - fixed length l_1 and l_2 during the time of motions C and D. These parameters are chosen so as to minimize the total time of motion T. The combination of optimal regimes with fixed length and simplest regimes of motion with variable length gives a possibility to transfer a pendulum to any given point in a vertical plane with damping oscillations.

APPLICATION

The optimal control laws suggested in this paper were employed for elaborating crane automatic control systems.

CONCLUSION

Different problems of minimal time motion of a pendulum attached to a rigid body are solved under various boundary conditions and contraints imposed on both the velocity and acceleration of a body. The conditions imply that there are no oscillations in the initial and final states of the system. The results were tested experimentally for cranes carrying swinging loads and can be applied to control of various mechanical systems.

REFERENCES

Anselmino, E., and T.M. Liebling (1967). Zeitoptimale Regelung der Bewegung einer haengenden Last zwischen zwei beliebigen Randpunkten. Proc. 5 th International Analogue Computation Meetings, Lausanne, Vol. 1, pp. 482-492.

Banichuk, N.V., and F.L. Chernousko (1975). The determination of optimal and suboptimal controls in one oscillating mechanical system. Izvestija of USSR Academy of Sciences. Mech. of Solids, 2, 39-43.

Beeston, J.W. (1967). Solution of the time-optimal control problem for systems of similar structure. Electron. Lett., Vol. 3, 378-379.

Chernousko, F.L. (1975). Optimal displacement of a pendulum. Prikl. Mat. & Mekh. (USSR), 5, 806-816.

Chernousko, F.L. (1977). Optimal control and dynamics of oscillating systems. Theoretical and Applied Mechanics. Proceedings of the 14 th IUTAM Congress Delft (1976), North-Holland Publishing Company, pp. 331-341.

Flügge-Lotz , I., and Mih Yin (1961). The optimum response of second-order, velocity-controlled systems with contactor control. Trans. ASME Ser. D, Vol. 83, pp. 59-64.

Flügge-Lotz, I., and H.A.Titus (1962). The optimum response of full third-order systems with contractor control. Trans. ASME Ser. D, Vol. 84, pp.554-558.

Hippe, P. (1970). Zeitoptimale Steuerung eines Erzentladers. Regelungstechnic (Germany), 18, H. 8, 346-350.

Mamaliga, V.M., and F.L. Chernousko (1977). The control by displacement of a load in vertical plane. Izvestija of USSR Academy of Sciences Mech. of Solids, 4, 93-101.

Mamaliga, V.M. (1979). The optimal control of some oscillating system. Izvestija of USSR Academy of Sciences. Mech. of Solids, 3, 8-17.

Mamaliga, V.M. (1979). The contstruction of time-optimal feedback control for oscillating system. Izvestija of USSR Academy of Sciences, Mech. of Solids. 3, 37-45.

Mikhailov, N.N., and G.A. Novoseltseva (1964) Optimal transient proceses in systems with prediction. Tekh. Kibern. (USSR), 1, 187-195.

Mikhailov, N.N., and G.A. Novoseltseva (1965). Optimal processes in third-order systems with complex poles. Avtom. & Telemekh. (USSR), 9,1502-1513

Moroz, A.I. (1969). Time-optimal feedback control for linear third-order systems, I, II, III. Avtom. & Telemekh. (USSR), 5,7,9, 5-19, 18-29, 5-15.

Pontryagin, L.S., V.G. Boltyansky, R.V. Gamkrelidzo, Ye. F. Mishchenko (1969) Mathematical Theory of Optimal Processes. Maila. ,pscpw. (in Russian).

Sokolov, B.N. (1977). Time-optimal start of a pendulous load with bounded both velocity and acceleration of equilibrium point, Izvestija of USSR Academy of Sciences, Mech. of Solids., 6,38-43.

Sokolov, B.N., and F.L.Chernousko (1976). About optimal displacement of a pendulous load. Izvestija of USSR Academy of Sciences. Mech. of Solids., 4, 26-33.

Sokolov, B.N., and F.L. Chernousko (1977). Optimal start of a pendulum. Izvestija of USSR Academy of Sciences. Mech. of Solids., 2, 18-23.

Troitskii, V.A. (1976). Optimal processes of oscillations of mechanical systems Mashinostroeniye, Leningrad.

Zaremba, A.T., and B.N. Sokolov (1978). About one problem of the optimal start of a pendulum with bounded velocity and integral criterion of a quality. Izvestija of USSR Academy of Sciences. Mech. of Solids., 4, 30-33.

Zaremba, A.T., and B.N.Sokolov (1979). About optimal combination of acceleration and braking of a pendulum suspension point during the start of a pendulous load. Izvestija of USSR Academy of Sciences. Mech. of Solids., 2, 18-22.

ON THE APPLICATION OF NONLINEAR PROGRAMMING TO THE SOLUTION OF OPTIMAL OUTPUT-CONSTRAINED REGULATOR PROBLEMS

J. R. Knox

Lockheed-Georgia Company, Marietta, Georgia 30063, USA

Abstract. Optimization techniques based on nonlinear programming are used to compute the constant, optimal output feedback gains, for linear multivariable control systems. The computation of these feedback gains provides a useful design tool in the development of aircraft active control systems. Broyden-Fletcher-Goldfarb-Shanno (BFGS), Davidon-Fletcher-Powell (DFP), and Newton methods are used in conjunction with appropriate starting values to compute the optimal gains; and a comparison of the effectiveness of the techniques is given. Also a modification of the DFP Method in which an analytical approximation of the inverse Hessian is used as a priming value is developed and evaluated. An example problem, the optimal control of a flexible aircraft, is used to evaluate the techniques. Results indicate that the methods provide an efficient and cost effective solution of the optimal output feedback problem.

Keywords. Nonlinear programming; numerical methods; optimization; optimal control; multivariable control systems.

INTRODUCTION

It is widely realized, that to attain the maximum benefits of actively controlled and control-configured aircraft, future flight control systems will require multiple inputs and controllers. The development of these systems will require the use of intermediate to large-scale models of the aircraft structure and control systems.

Modern control theory provides a conceptual and theoretical tool for the design of these systems, especially in context of Linear-Quadratic Optimization. If all of the system states are sensed (state feedback), efficient and convergent algorithms are available to solve the time-invariant Linear-Quadratic-Regulator problem (Kleinman, 1968). However, it is not practical to sense or estimate all of the large number of system states required for a reasonably accurate aircraft model. Hence, for these models the available feedback is a linear combination of the sensed states and is referred to as output feedback.

The optimal output-constrained regulator problem was formulated by Axsäter (1966), and Levine and Athans (1970), who also provided algorithms which had the property of monotonically decreasing the criterion function at each step, hence ensuring convergence for starting values sufficiently close to a finite solution. Unfortunately these methods required the solution of a computationally expensive nonlinear matrix equation at each iteration, which caused them to be impractical except in the solution of very small systems. In an effort to avoid this complication, variant algorithms requiring only the solution of linear matrix equations were proposed by Kurtarian and Sidar (1974), and Anderson and Moore (1971). However, these methods do not guarantee descent, and Söderström (1978) has shown nonconvergence for a simple, nonpathological counter-example.

The solution of the Optimal Output-Constrained Regulator problem is further complicated by the fact that multiple minima may exist, hence any method may converge to different solutions, depending on the starting values used. The utilization of several randomly distributed starting values has been suggested as a technique for finding a global minimum. Also methods have been suggested in which the output constrained solution is obtained by solving a sequence of problems with increasingly severe constraints, starting from the unconstrained (state feedback) solution.

In the approach used in this paper the starting value is directly obtained from the state feedback solution by a technique suggested by Petkovski and Rakic (1978).

The utilization of nonlinear programming in the solution of the Output-Constrained Regulator problem has been considered by several authors. Choi and Sirisena (1974) used the Davidon-Fletcher-Powell method; and Horisberger and Belanger (1974) used a conjugate gradient method to solve example problems. These methods require only gradient or first derivative information and provide, in theory, monotonic descent of the cost function.

For starting values sufficiently close to the solution, Newton's method may provide a more effective solution than gradient techniques. However, it requires the costly computation of second derivatives, and may not converge if the starting value is far from the solution. Stein and Henke (1971) derived analytic expressions for the second derivatives and used Newton's method to solve aircraft flight control problems involving multiple plants. In an effort to ensure convergence of Newton's method, and also to obtain an absolute minima, they used the state feedback solution as a starting value and approached the output constrained solution through a computationally expensive parameterization procedure. A similar technique based on an approximation of the second derivatives was presented by Vandierendonck (1972). Also Bingulac, Cuk, and Calovic (1975) present a rather involved technique of approaching the output-constrained solution from the state feedback solution.

This paper presents a discussion of the efficient utilization of nonlinear programming in the solution of the Optimal Output-Constrained Regulator problem. Starting values which experience has shown to be close to the global solution point are directly computed, hence providing a more efficient solution than prior methods. Also, a modification of the DFP method is presented which increases its effectiveness and provides monotonic descent. An example problem is used to compare the effectiveness of this method with the DFP, BFGS, and Newton methods. Results show that the methods used provide a computationally cost effective means of computing the optimal gains.

PROBLEM DEFINITION

Aeroelastic aircraft models may be approximated in the form of matrix polynomials in the Laplacian operator s. Using appropriate techniques these polynomials may in turn be realized in state space form, e.g. see Newsome (1978), Roger (1977).

Hence, a general aircraft model may be defined as

$$\frac{dz}{dt} = \dot{z}(t) = D\,z(t) + E\,u(t) \qquad (1)$$

$$y(t) = G\,z(t)$$

with z, u, and y, n, m, and p vectors respectively; and D an n by n matrix, E an n by m matrix and G a p by n matrix.

For computational efficiency we may without loss of generality use the linear transformation, $x = Tz$;

$$T = \left[\begin{array}{c|c} \multicolumn{2}{c}{G_{pn}} \\ \hline 0_{n-p,p} & I_{n-p,n-p} \end{array}\right]_{n,n} \qquad (2)$$

to provide a system model equivalent to (1), i.e.

$$\dot{x} = [T\,D\,T^{-1}]x + [T\,E]u$$

$$y = [I_{p,p} \mid 0_{p,n-p}]\,x \quad \text{or equivalently} \qquad (3)$$

($I_{p,p}$ is the p by p identity matrix)

$$\dot{x} \equiv Ax + Bu$$

$$y \equiv Cx$$

The optimal control (Levine and Athans, 1970) is obtained by minimizing the expected value of the standard quadratic cost function, under the assumption that the initial states are randomly distributed, i.e. minimize the scalar performance index,

$$J(F) = E\left\{\int_0^\infty [x'Qx + u'\,Ru]\,dt\right\} \qquad (4)$$

subject to

$$\dot{x}(t) = Ax(t) + Bu(t) \quad x(0) = x_o \qquad E\{x_o\,x_o'\} = X_o$$

$$y(t) = Cx(t) \qquad u(t) = Lx(t) \quad L \equiv FC \qquad (5)$$

with $E(\cdot)$ the expectation operator, x' the transpose of x, and R, Q, and X_o suitably chosen matrix design parameters, e.g. see Levine, Johnson, and Athans (1971); Athans (1971).

From the results of Stein and Henke (1971)

$$J(F) = TR\{[Q + L'RL]\,X\} = TR\,\{VX_o\} \qquad (6)$$

$$\frac{\partial J}{\partial F_{ij}} = 2TR\left\{\left[L'R - VB\right]\left[\frac{\partial L}{\partial F_{ij}}\right]X\right\} \qquad (7)$$

$$\frac{\partial^2 J}{\partial F_{rs}\partial F_{ij}} = 2TR\left\{\left[\frac{\partial L}{\partial F_{rs}}\right]' R\left[\frac{\partial L}{\partial F_{ij}}\right]X \right. \qquad (8)$$

$$+ \left[L'R - VB\right]\left[\frac{\partial L}{\partial F_{ij}}\right]\left[\frac{\partial X}{\partial F_{rs}}\right]$$

$$\left. + \left[L'R - VB\right]\left[\frac{\partial L}{\partial F_{rs}}\right]\left[\frac{\partial X}{\partial F_{ij}}\right]\right\}$$

where $L = FC$, (L is an implicit function of F)

$X_o = E\,\{x_o\,x_o'\} \equiv$ state covariance matrix,

$TR\,(\cdot)$ Is the trace operator; and

V, X, and $\partial X/\partial F_{ij}$ are solutions of the matrix algebraic equations,

$$V[A-BL] + [A-BL]' V + Q + L'RL = 0 \quad (9a)$$

$$X[A-BL]' + [A-BL] X + X_o = 0 \quad (9b)$$

$$\left[A-BL\right]\left[\frac{\partial X}{\partial F_{ij}}\right] + \left[\frac{\partial X}{\partial F_{ij}}\right]\left[A-BL\right]' + \left[B\frac{\partial L}{\partial F_{ij}}\right]X + X\left[B\frac{\partial F}{\partial F_{ij}}\right]' = 0 \quad (9c)$$

The optimal control is given by $u(t) = R^{-1}B'VXC' [CXC']^{-1} x(t)$. Under the transformation given by equation (2), $\partial L/\partial F_{ij}$ assumes a particularly simple form, and after some simplification one obtains

$$\frac{\partial J}{\partial F_{ij}} = 2 \sum_{q=1}^{n} X_{qj} [RL - B'V]_{iq} \quad (10)$$

where $[RL - B'V]_{iq}$ indicates the iq^{th} element of the matrix $[RL-B'V]$, and

$$\frac{\partial^2 J}{\partial F_{rs}\partial F_{ij}} = 2 R_{ri} X_{js} + 2 \sum_{q=1}^{n} [RL - B'V]_{iq} \left[\frac{\partial X}{\partial F_{rs}}\right]_{jq} + 2 \sum_{q=1}^{n} [RL - B'V]_{rq} \left[\frac{\partial X}{\partial F_{ij}}\right]_{sq} \quad (11)$$

DISCUSSION OF SOLUTION METHODOLOGY

The problem may be placed in the context of nonlinear programming by considering J as the scalar function of an mp vector, f, composed of elements of the feedback matrix F as

$$f \triangleq \begin{bmatrix} F_{11} \\ F_{12} \\ \vdots \\ F_{1p} \\ F_{21} \\ F_{22} \\ \vdots \\ F_{2p} \\ \vdots \\ F_{mp} \end{bmatrix} \quad (12)$$

For the problem to be defined f must be a stabilizing feedback, i.e. the system under the control of f must be stable. Hence the problem may be solved by minimizing J (f) subject to the system constraints, equations (5), and to the condition that [A-BFC] be a stability matrix. Since first and second partial derivative information is computable from equations (10) and (11), both Newton and Quasi-Newton methods may be used to implement a solution.

The well-known Davidon-Fletcher-Powell (Fletcher and Powell, 1963), Broyden-Fletcher-Goldfarb-Shanno (Rauch, 1978), and Newton (Luenberger, 1973) methods were implemented along with a modification of the Davidon-Fletcher-Powell method, and were used to solve an example problem, the optimal control of a flexible aircraft. A discussion of specific aspects of the programming techniques, and an evaluation of the effectiveness of the methods in solving the example problem are presented in the following sections.

Solution of Covariance Equations

Computation of the first and second derivatives requires solution of the matrix covariance equations [equations (9)]. Algorithms have been developed to provide an efficient solution of these equations. The method used here is due to Smith (1968). Also a method developed by Bartels and Stewart (1972) has been used successfully.

Modified Davidon-Fletcher-Powell Method

Using only the first term of equation (11), an approximation of the Hessian of J with respect to the vector f may be obtained. It may be shown that this matrix is the Kronecker product of the positive definite matrices R and X (Knox and McCarty, 1978) and hence is positive definitive. The inverse of this matrix may be used to restart the DFP method every N_o steps. The usefulness of this method depends upon how closely the Hessian is approximated. Because the inverse Hessian is approximated by the restart value, it is assumed that the DFP method will provide an adequate, more exact approximation in less than mp (the order of the Hessian) steps, hence the arbitrary parameter N_o is chosen less than mp.

The effectiveness of this technique must be demonstrated by numerical experiment, as it is difficult to analytically determine. An interesting observation is given by the fact that if all the system states are sensed the solution is given by

$$F \equiv L^* = R^{-1}B'V \quad \text{or} \quad RL^* - B'V = 0 \quad (13)$$

In this case the second and third terms of equation (11) become zero, and the exact, positive definite Hessian of J (f) may be computed using only the first term of equation (11).

The computational cost of using this approximation is small compared to the exact Hessian. The comparative cost is illustrated by the fact that computation of the exact Hessian at each iteration or line search requires mp covariance equation solutions in excess of the two required to compute the gradient of J with respect to f, while the approximation of the Hessian requires no additional covariance equation solutions.

Starting Values

Two starting values of F were used to initiate the algorithms, both based on the solution of the corresponding unconstrained (state feedback) problem. The first is given by

$$F_o = L^* X^* C' \, [C \, X^* C']^{-1} \tag{14}$$

where L^* solves the state feedback problem and X^* is the corresponding solution of the covariance equation (9b). This solution provides the minimum excitation error between the control F_o and the control L^* for the system given by equation (5), (Kosut, 1970). The second starting value used is given by

$$F_1 = L^* C' \, [C \, C']^{-1} \tag{15}$$

This is the minimum norm solution (Kosut, 1970). The starting value provided by equation (14) has been found especially useful in practice; however, there is no guarantee that in the general case either equation (14) or equation (15) will provide the requisite stabilizing feedback.

Line Search

The cost function is reduced to a scalar function of a real variable α for the line search portion of the optimization process, e.g. see Luenberger (1973). The objective of the line search is to minimize the cost function J with respect to α. The specific method used is described as follows. An α, say α_k, is found such that α_k is greater than the minimizing α. This is accomplished by computing a sequence of α's as follows. First an initial α is chosen as

$$\alpha_o = \min\{[J(\alpha)/(dJ/d\alpha)]\big|_{\alpha=0}, 1.\} \tag{16}$$

Successive values are given by the formula

$$\alpha_j = \alpha_{j-1}(2)^j \qquad j = 1, 2, \ldots \tag{17}$$

until an $\alpha_j = \alpha_k$ is found such that α_k is greater than the minimizing α.

In order for the line search to be effective, provision must be made for the possibility that the search sequence may carry the system into an unstable region, in which case J is not defined. The search is restricted to the region of stability by the following technique. If an α_j, say α_i, is encountered that produces an unstable system as evidenced by nonconvergence of the algorithm used to solve the covariance equations [equations (9)], the search sequence [equation (17)] is restarted using $\alpha_o = \frac{1}{4}\alpha_i$ as the initial value.

Once an α_k is found, cubic interpolation, using values of the cost function and its derivative at the end points of the interval $[\alpha_o, \alpha_k]$ is used to compute an approximate minimizing α, say α_m. Then a check is made to determine if the cost function evaluated at α_m is less than its value at the end points of the interval. If this requirement is met, the line search is considered finished. If this is not the case, the intervals $[\alpha_o,\alpha_m]$, and $[\alpha_m,\alpha_k]$ are checked to determine which one contains the minimum. Then the cubic interpolation procedure is utilized in conjunction with the appropriate interval to compute a new value of α_m, and the process repeated until a suitable α_m is found. Double precision arithmetic (16 decimal digits) was used for the line search computations.

EXAMPLE PROBLEM

The control of a transport aircraft was used as an example problem. A seventh-order model of the C-141A aircraft was obtained by reduction techniques from a larger, more exact representation of the aeroelastic dynamics. Five of the seven system states were sensed, specifically, normal accelerations at the wing-tip, mid-wing, and the center of gravity; and the aircraft's attitude and attitude rate at the center of gravity.

The DFP, BFGS, Newton, and previously defined Modified DFP methods were used to compute the optimal gains. Convergence was assumed in every case if

$$\left[\frac{\partial J}{\partial f}\right]' \left[\frac{\partial J}{\partial f}\right] < .001$$

where $\partial J/\partial f$ is the gradient of J with respect to f. The techniques were programmed in Fortran on the Lockheed-Georgia Univac 1100/11 computer system.

The criteria used to evaluate the methods are the Central Processor Unit (CPU) time, the number of covariance equations, and the number of line searches required for convergence.

All the methods converged to the same solution point if convergence was achieved. A comparison of the methods is given in Table 1.

CONCLUSION

The results of Table 1 show that the optimization techniques provided an effective method of computing optimal control gains for the example problem. A solution was obtained in 50 seconds of Central Processor Unit (CPU) time on the relatively slow Univac 1100/11 computer. The minimum error excitation starting value was shown to be superior to the minimum norm. The value of the cost function corresponding to the minimum error

TABLE 1 Comparison of Methods

Method	Starting Value	CPU (Seconds)	Covariance Eq. Solutions Req'd.	No. of Line Searches	Restart Parameter N_0
Modified DFP	Eq. (14)	50	68	9	4
Newton	Eq. (14)	57	78	4	--
BFGS	Eq. (14)	117	164	23	--
DFP	Eq. (14)	136	186	23	--
Modified DFP	Eq. (15)	85	114	16	4
Newton	Eq. (15)	Failed to Converge			
BFGS	Eq. (15)	155	216	33	--
DFP	Eq. (15)	287	390	44	--

excitation starting value was very close to the globally optimal value of the corresponding state feedback problem. This fact in conjunction with the observation that Newton's method converged in only four iterations provides an indication of the closeness of this starting value to the globally optimal solution.

However, there is no guarantee that the minimum error excitation starting value will provide the positive definite Hessian matrix required for utilization of Newton's method. A possible alternative approach is to use the Modified DFP or BFGS methods initially in the optimization process, then switch to Newton's method as the solution is approached. Also there is no guarantee that this or any other numerical technique will converge to the globally optimal solution.

The results also show that increased efficiency is obtained by computation of second derivative information near the solution point, as evidenced by the clear superiority of the Modified DFP and Newton methods over the BFGS and DFP methods when the minimum excitation error starting value was used. Newton's method failed to converge for the minimum norm starting value. With this exception the procedures converged to the same solution in every case. Convergence properties of these methods are discussed in Luenberger (1973). It is also noted that the BFGS method was superior to the DFP for both starting values.

A computational evaluation of the methods presented is dependent upon the size and conditioning of the problem, the ratio of the number of sensed states to the number of states, and the line search method. The importance of the search technique is evidenced by the fact that it was necessary to use double precision arithmetic (16 decimal digits) for the line search computations in order to achieve convergence. For these reasons it is somewhat difficult to generalize the results of this study to larger, more general systems.

However, two observations indicate the utility of these methods in the solution of large scale systems. First, although the example problem was relatively small, it was ill-conditioned, with the condition number, r, of the Hessian equal to 428754.11 at the starting value of equation (14) and equal to 819573.84 at convergence (r being the ratio of largest to smallest eigenvalue of the Hessian at the point in question). Secondly, the algorithm used to solve the covariance equations (Smith, 1968) has been successfully applied to large scale problems (Smith, 1971).

Hence, in summary, the results indicate that nonlinear programming methods when used in conjunction with appropriate starting values provide an effective and efficient method of computing the optimal, constant output feedback gains for active aircraft control systems. Also, it was observed that a simple modification of the DFP method will improve its efficiency in the solution of this problem.

Related Problems

There are several variants and extensions of this problem that possibly may be solved by nonlinear programming. Johnson and Athans (1970) formulated the optimal dynamic output compensation problem in the form of equation (6), but with other interpretations of involved matrices. The computation of the Optimal Constrained Output Feedback gains of stochastic linear continuous systems was considered by Basuthakur and Knapp (1975); the case of linear discrete systems including the stochastic case was considered by Elmer and Vandelinde (1973). The inclusion of additional feedback constraints in order to realize certain system criteria was considered in the context of constrained optimization by Knox and McCarty (1978).

REFERENCES

Anderson, B. D. O., and J. B. Moore (1971). *Linear Optimal Control*. Prentice Hall, Englewood Cliffs, New Jersey.

Athans, M. (1971). The role and use of the Stochastic Linear-Quadratic-Gaussian problem in control system design. *IEEE Trans. Autom. Control*, AC-16, 529-552.

Axsater, S. (1966). Sub-optimal time-variable feedback control of linear dynamic systems with random inputs. *Int. J. Control*, 4, No. 6, 549-566.

Bartels, R. H., and G. W. Stewart (1972). Algorithm 432 - Solution of the matrix equation AX + XB = C. *Commu. ACM*, 15, No. 9, 820-826.

Basuthakur, S., and C. H. Knapp (1975). Optimal constant controllers for stochastic linear systems. *IEEE Trans. Autom. Control*, AC-20, 664-666.

Bingulac, S. P., N. M. Cuk, and M. S. Calovic (1975). Calculation of optimum feedback gains for output-constrained regulators. *IEEE Trans. Autom. Control*, AC-20, 164-166.

Choi, S. S., and S. R. Sirisena (1974). Computation of optimal output feedback gains for linear multivariable systems. *IEEE Trans. on Autom. Control*, AC-19, 257-258.

Ermer, C. M., and V. D. Vandeline (1973). Output feedback gains for a linear-discrete stochastic control problem. *IEEE Trans. Autom. Control*, AC-18, 154-157.

Fletcher, R., and M. J. D. Powell (1963). A rapidly convergent descent method for minimization. *Comput. J.*, 6, 163-168.

Horisberger, H. P., and R. R. Belanger (1974). Solution of the optimal constant output feedback problem by conjugate gradients. *IEEE Trans. Autom. Control*, AC-19, 434-435.

Johnson, T. L., and M. Athans (1970). On the design of optimal constrained dynamic compensators for linear constant systems. *IEEE Trans. Autom. Control*, AC-15, 658-660.

Kleinman, D. L. (1968). On an iterative technique for Riccati equation computations. *IEEE Trans. Autom. Control*, AC-13, 114-115.

Knox, J. R., and J. M. McCarty (1978). Algorithms for computation of optimal constrained output feedback for linear multivariable systems. *AIAA Paper 78-1290, AIAA Guidance and Control Conference*, Palo Alto, Ca.

Kosut, R. L. (1970). Suboptimal control of linear time-invariant systems subject to control structure constraints. *IEEE Trans. Autom. Control*, AC-15, 557-563.

Kurtarian, B.-Z., and M. Sidar (1974). Optimal instantaneous output-feedback controllers for linear stochastic systems. *Int. J. Control*, 19, 797-816.

Levine, W. S., and M. Athans (1970). On the determination of the optimal constant output feedback gains for linear multivariable systems. *IEEE Trans. Autom. Control*, AC-15, 44-48.

Levine, W. S., T. L. Johnson, and M. Athans (1971). Optimal limited state variable feedback controllers for linear systems. *IEEE Trans. Autom. Control*, AC-16, 785-792.

Luenberger, D. G. (1973). *Introduction to Linear and Nonlinear Programming*. Addison-Wesley Publishing Co., Reading, Mass.

Newsome, J. R. (1978). Synthesis of actual flutter suppression systems using control theory. *AIAA Paper 78-1270, AIAA Guidance and Control Conference*, Palo Alto, Ca.

Petkovski, D. J. B., and M. Rakic (1978). On the calculation of optimum feedback gains for output-constrained regulators. *IEEE Trans. Autom. Control*, AC-23, 760.

Rauch, H. E. (1978). Problems and techniques in constrained optimization. *AIAA Paper 78-1408, AIAA Guidance and Control Conference*, Palo Alto, Ca.

Roger, K. (1977). Airplane math modelling methods for active control design. *AGARD Conference Proceedings No. 228, Structural Aspects of Active Controls, AGARD CP-228.*

Smith, P. G. (1971). Numerical solution of matrix equation AX + XA' + B = 0. *IEEE Trans. Autom. Control*, AC-16, 278-279.

Smith, R. A. (1968). Matrix equation XA + BX = C. *SIAM J. Appl. Math.*, 16 198-201.

Söderström, T. (1978). On some algorithms for design of optimal constrained regulators. *IEEE Trans. Autom. Control*, AC-23, 1100-1101.

Stein, G., and A. H. Henke (1971). A design procedure and handling quality criteria for lateral-directional flight control systems. *Air Force Flight Dynamics Laboratory TR-70-152*, Wright-Patterson Air Force Base, Ohio.

Vandierendonck, A. J. (1972). Design method for fully augmented system for variable flight conditions. *Air Force Flight Dynamics Laboratory TR-71-152*, Wright-Patterson Air Force Base, Ohio.

IDENTIFICATION BY A COMBINED SMOOTHING NONLINEAR PROGRAMMING ALGORITHM

A. E. Bryson, Jr.* and A. B. Cox**

**Paul Pigott Professor of Aeronautics and Astronautics, Stanford University, Stanford, California, USA*
***Department Staff Engineer, TRW Defense and Space Systems Group, Redondo Beach, California, USA*

Abstract. An offline algorithm is developed for identification of parameters of linear, stationary, discrete, dynamic systems with known control inputs and subjected to process and measurement noise with known statistics. Results of the algorithm include estimates of the parameters and smoothed estimates of the state and process noise sequences. The problem is stated as the minimization of a quadratic performance index. This minimization problem is then converted to a nonlinear programming problem for determining the optimum parameter estimates. The new algorithm is shown to be cost competitive with the currently popular filtering-sensitivity function method. A third order example with simulated data is presented for comparison.

Keywords. Identification; nonlinear programming; smoothing; linear systems; discrete systems; iterative methods.

NOMENCLATURE

a	Vector of uncertain parameters (pX1)
A	A priori covariance matrix for uncertain parameters (pXp)
B	Control effectiveness matrix (nXc)
$\bar{B}$	Control effectiveness matrix in modal form (nXc)
c	Number of control input sequences u
H	State multiplier matrix for determining measurements (mXn)
I	Identity matrix
J	Quadratic performance index (scalar)
$\bar{J}$	Augmented quadratic performance index (scalar)
k	Current discrete time step
m	Number of measurements
n	Number of system states
N	Number of discrete time steps in data span
p	Number of uncertain parameters in the system model
P( )	Covariance matrix of states (nXn)
q( )	Vehicle pitch rate
Q	Covariance matrix of process noise sequence (scalar)
R	Covariance matrix of measurement noise sequence (mXm)
T	Transformation matrix to diagonalize transition matrix of two point boundary value problem (nXn)
u( )	Input control vector sequence (cX1)
v( )	Gauss markov purely random noise vector sequence (mX1)
w( )	Gauss markov purely random process noise sequence
$\bar{w}$	Mean value of w( )
x( )	State vector sequence (nX1)
$\hat{x}$( )	Smoothed estimate of state vector sequence (nX1)
z( )	Measurement sequence (mX1)
Z	Block diagonal matrix of eigenvalues (2nX2n)
Z_1	Block diagonal submatrix of Z containing stable eigenvalues (nXn)
Z_2	Block diagonal submatrix of Z containing unstable eigenvalues (nXn)
α	Matrix of rank one used to update inverse curvature matrix
α( )	Angle of attack in example (rad)
γ_i	Components of process noise effectiveness vector (Γ)
Γ	Process noise effectiveness vector (nX1)
η( )	Wind gust process noise in third order example
λ( )	Lagrange multiplier vector sequence (nX1)
ξ_1( ), ξ_2( )	Modal state vector sequences (nX1)
Υ	Current best estimate of J_{aa}^{-1}
ϕ_{ij}	Components of the transition matrix Φ
Φ	Transition matrix (nXn)

INTRODUCTION

Although smoothing methods have been suggested for use as parameter estimators for many years (Cox, 1964), few competitive smoothing algorithms have been developed for simultaneous parameter identification and state estimation of linear, stationary, dynamic systems. Current research in offline smoothing methods has been directed toward nonlinear systems (Bach, 1977; Chang, 1977).

This paper presents a combined state estimation/parameter identification procedure for linear, stationary, discrete, dynamic systems. This procedure is an application of smoothing to parameter identification and is competitive in cost with procedures currently used. Process and measurement noise sequences in the system are approximated by gauss markov purely random sequences. The algorithm is iterative and for offline use.

A NEW SMOOTHING-IDENTIFICATION ALGORITHM

The parameter identification problem for linear, stationary, discrete, dynamic systems may be stated as follows:

- Given

  1. The sequences x(k), w(k), v(k), z(k), and u(k) satisfy model equations; i.e., Eqs. (1) and (2).

$$x(k+1) = \Phi(a)x(k) + B(a)u(k) + \Gamma(a)w(k) \quad k=0, 1, \ldots, N\text{-}1 \tag{1}$$

$$z(k) = H(a)x(k) + v(k) \quad k=1, 2, \ldots, N\text{-}1 \tag{2}$$

  2. Φ, B, Γ, and H are constant matrices whose elements are known functions of an uncertain parameter vector a.

  3. The vector a is random with mean value a_0 and covariance A.

  4. The measurement sequence z(k) is known over some span, k=1, 2, . . . , N-1.

  5. The control input sequence u(k) is known over some span, k=0, 1, . . . , N-1.

  6. The sequences w(k) and v(k) are gauss-markov purely random sequences with mean value $\bar{w}$ and zero, respectively, and covariances Q and R, respectively.

  7. The initial and final states of the system, x(0), and x(N), are random vectors with mean values x_0 and x_f, respectively, and covariances P_0 and P_f, respectively.

- Find

  The sequences x(k) and w(k) and the vectors a, x(0), and x(N) which minimize the performance index.

$$\begin{aligned} J = {} & 0.5\,[a-a_0]^T A^{-1}[a-a_0] \\ & + 0.5\,[x(0)-x_0]^T P_0^{-1}[x(0)-x_0] \\ & + 0.5\,[x(N)-x_f]^T P_f^{-1}[x(N)-x_f] \\ & + 0.5 \sum_{k=1}^{N-1} [z(k)-Hx(k)]^T R^{-1}[z(k)-Hx(k)] \\ & + 0.5 \sum_{k=1}^{N-1} [w(k)-\bar{w}]^T Q^{-1}[w(k)-\bar{w}] \end{aligned} \tag{3}$$

Conversion to a Nonlinear Programming Problem

The stated problem can be converted to a nonlinear programming problem by adjoining the constraint equation, Eq. (1), to the performance index using the Lagrange multiplier sequence $\lambda(k+1)$. Thus,

$$\bar{J} = J + \sum_{k=0}^{N-1} \lambda^T(k+1)\,[\Phi(a)x(k) + B(a)u(k) + \Gamma(a)w(k) - x(k+1)] \tag{4}$$

Using standard procedures (Bryson and Ho, 1975; Cox, 1979), necessary conditions for a stationary value of $\bar{J}$ (and hence J) are

$$x(k+1) = \Phi(a)x(k) + B(a)u(k) + \Gamma(a)w(k) \tag{5}$$

$$\lambda(k) = \Phi^T(a)\lambda(k+1) - H^T(a)R^{-1}[z(k)-H(a)x(k)] \tag{6}$$

$$w(k) = \bar{w} - Q\Gamma^T(a)\lambda(k+1) \tag{7}$$

$$x(0) = x_0 - P_0\Phi^T(a)\lambda(1) \tag{8}$$

$$x(N) = x_f + P_f\lambda(N) \tag{9}$$

and

$$\bar{J}_a\big|_{a=\hat{a}} = 0 \tag{10}$$

These necessary conditions form a linear two point boundary value problem (TPBVP) for determining w(k) and x(k) given the uncertain parameter vector a that is combined with a nonlinear programming problem for determining the optimal parameter vector $\hat{a}$. The algorithms presented are based on finding x(k), x(0), and a that cause these necessary conditions to be satisifed. Solutions for the linear TPBVP are found successively using different values of $\hat{a}$ ($\hat{a}_0$, $\hat{a}_1$, . . .). These solutions are used to determine $\bar{J}_a|_{\hat{a}=a_i}$. A nonlinear programming technique is then used to make an improved estimate ($\hat{a}_{i+1}$) that will decrease the magnitude of $\bar{J}_a|_{a=\hat{a}_{i+1}}$ on the next iteration.

A Modal Method for Solving the Two Point Boundary Value Problem

Evaluation of $\bar{J}_a$ requires solution of the linear TPBVP given by Eqs. (5) through (9) using the best estimate of the uncertain parameter vector, a [Eq. (23) and Cox, 1979]. The resulting state estimate sequence, $\hat{x}(k)$, is known as the smoothed estimate of the state sequence. The currently accepted method for solving this TPBVP involves the solution of a matrix Smoothing-Riccati equation in either a forward or backward sweep through the measurement sequence (Bryson and Ho, 1975). A significant reduction in computations can be achieved by transforming the TPBVP into modal coordinates. For long data spans, this modal method results in a significant reduction in the computations required to determine $\bar{J}_a$.

Briefly, the modal method utilizes the linearity of the TPBVP to advantage by transforming the system of equations in modal form. The TPBVP can be expressed as

$$\begin{bmatrix} \hat{x}(k+1) \\ \lambda(k+1) \end{bmatrix} = \bar{\Phi} \begin{bmatrix} \hat{x}(k) \\ \lambda(k) \end{bmatrix} + \begin{bmatrix} B \\ 0 \end{bmatrix} u(k) + \begin{bmatrix} \Gamma \\ 0 \end{bmatrix} \bar{w} \tag{11}$$

$\hat{x}(0)$ and $\hat{x}(w)$ specified.

By utilizing an appropriate similarity transformation, the transition matrix can be diagonalized; i.e.,

$$Z = T^{-1}\bar{\Phi}T = \begin{bmatrix} Z_1 & 0 \\ 0 & Z_2 \end{bmatrix} \tag{12}$$

where T is the eigenvector matrix of $\bar{\Phi}$ and Z is a block diagonal eigenvalue matrix. The eigenvalues of $\bar{\Phi}$ occur in pairs which are reciprocals of each other. Therefore, half of the eigenvalues are stable in a forward direction in time, and half are stable in a backward direction in time. This symmetry can be used to advantage in solving the TPBVP. By appropriate ordering of the columns of T, the Z matrix can be partitioned into a stable forward matrix, Z_1, and a stable backward matrix, Z_2. Transformation of the system to modal form then yields

$$\begin{bmatrix} \xi_1(k+1) \\ \xi_2(k+1) \end{bmatrix} = \begin{bmatrix} Z_1 & 0 \\ 0 & Z_2 \end{bmatrix} \begin{bmatrix} \xi_1(k) \\ \xi_2(k) \end{bmatrix} + [\bar{B}]u(k) + [\bar{\Gamma}]\bar{w} \quad (13)$$

$$C_1\xi_1(1) + D_1\xi_2(1) = E_1 \quad (14a)$$

$$C_2\xi_1(N) + D_2\xi_2(N) = E_2 \quad (14b)$$

which represents two sets of decoupled first and second order equations with coupled boundary conditions. The first set, $\xi_1(k)$, is stable in the forward direction in time; the second set, $\xi_2(k)$, is stable in the backward direction of time. This simplification in the coupling between state variables reduces computation requirements significantly. Each real eigenvalue produces a first order state equation; each complex pair of eigenvalues yields a second order state equation.

One can now capitalize on the decoupling and linearity of the equations to determine a solution to the TPBVP. Each state equation set can be expressed as the sum of a homogeneous and a particular solution; i.e.,

$$\xi_i(k) = \xi_i^{(p)}(k) + \xi_i^{(h)}(k) \quad i = 1, 2 \quad (15)$$

The particular solution is chosen to satisfy

$$\xi_1^{(p)}(1) = 0 \quad (16)$$

$$\xi_2^{(p)}(N) = 0 \quad (17)$$

Then,

$$\xi_1^{(p)}(k+1) = Z_1\xi_1^{(p)}(k) + \bar{B}_1u(k) + \bar{\Gamma}_1\bar{w} \quad (18)$$

$$\xi_2^{(p)}(k+1) = Z_2\xi_1^{(p)}(k) + \bar{B}_2u(k) + \bar{\Gamma}_2\bar{w} \quad (19)$$

The homogeneous solutions will be of the form

$$\xi_1^{(h)}(k) = Z_1^k\xi_0 \quad (20)$$

$$\xi_2^{(h)}(k) = Z_2^{k-N}\xi_f \quad (21)$$

Substitution of Eqs. (15), (20), and (21) into the boundary conditions yields a set of 2n linear equations to solve for ξ_0 and ξ_f. With the homogeneous boundary conditions available, the homogeneous solutions can be obtained and added to the particular solutions to determine $\xi_1(k)$ and $\xi_2(k)$. The inverse similarity transformation can then be performed to obtain the desired sequences, $\widehat{x}(k)$ and $\lambda(k)$, and finally $\bar{J}_a$.

$$\begin{bmatrix} \widehat{x}(k) \\ \lambda(k) \end{bmatrix} = T \begin{bmatrix} \xi_1(k) \\ \xi_2(k) \end{bmatrix} \quad (22)$$

Matrix calculations above second order are not required to determine particular solutions, and matrix Riccati equations do not have to be "integrated". The major cost associated with the smoothing-modal method is the initial diagonalization process. In many cases, the computational time saved during the "integration" exceeds the time required for the initial diagonalization.

Determination of $\bar{J}_a$ While Solving the Two Point Boundary Value Problem

The gradient of the performance index with respect to the parameter vector, $\bar{J}_a$, can be found by a simple vector quadrature using the TPBVP solution (Cox, 1979); i.e.,

$$\frac{\partial \bar{J}}{\partial_{a(i)}} = \sum_{j=1}^{P} [a(j)-a_0(j)]/A(i,j)$$

$$-\sum_{k=1}^{N-1} [z(k)-Hx(k)]R^{-1}\frac{\partial H}{\partial a(i)}x(k)$$

$$+\sum_{k=1}^{N-1} \lambda^T(k+1)\left[\frac{\partial \Phi}{\partial a(i)}x(k) + \frac{\partial B}{\partial a(i)}u(k)\right.$$

$$\left.+\frac{\partial \Gamma}{\partial a(i)}w(k)\right] \quad (23)$$

for $i-1, 2, \ldots, p$

Parameter Estimation

After obtaining $\bar{J}_a$ for the current best estimate of the parameter a_i, the next step in the minimization procedure is to improve the parameter estimates. A modified Newton's method was chosen to minimize $\bar{J}$ with respect to the parameter vector a (Luenberger, 1965). This method assumes that $\bar{J}_a(a+\Delta a)$ can be approximated locally as a linear function of Δa, wherein

$$\bar{J}^T(a+\Delta a) \simeq \bar{J}_a^T(a) + \bar{J}_{aa}(a)\Delta a \quad (24)$$

At the minimum of J(a), $\bar{J}_a^T = 0$. Hence, the difference between the current value of a and the minimizing value is

$$\Delta a \simeq -\bar{J}_{aa}^{-1}\bar{J}_a^T(a) \quad (25)$$

i.e.,

$$a_{i+1} \simeq a_i - \bar{J}_{aa}^{-1}(a_i)\bar{J}_a^T(a_i) \quad (26)$$

The rank one correction procedure is used to estimate $\bar{J}_{aa}$. This method uses successive values of $\bar{J}_a$ to estimate $\bar{J}_{aa}^{-1}$ (Luenberger, 1965). Thus, if Υ is the current best estimate of the inverse curvature matrix, J_{aa}^{-1}, then an improved estimate is

$$\Upsilon = \Upsilon + \frac{[\Delta a - \Upsilon\Delta\bar{J}_1^T]\,[\Delta a - \Upsilon\Delta\bar{J}_a^T]^T}{\Delta J_a\,[\Delta a - \Upsilon\Delta\bar{J}_a^T]} \quad (27)$$

For the situation in which a is a scalar, Eq. (27) reduces to

$$\Upsilon \simeq \frac{\Delta a}{\Delta J_a} \quad (28)$$

The rank one correction method utilizes information from the change in the gradient as a function of parameter change and thus requires an initialization routine. A good initial estimate of the curvature matrix can be derived from gradient information by using orthogonal steps in the parameter vector space, i.e., finding the gradient with respect to each parameter individually. This approach facilitates the use of the rank one correction procedure in the search for a good approximation to the local inverse curvature matrix. The step size for the orthogonal initialization procedure is determined by the a priori covariance matrix, A.

The estimates for the gradient and inverse curvature matrix and Eq. (26) are used to improve the estimate of the parameter vector, a. The new estimate is then used to obtain an improved solution to the TPBVP, and the iterative procedure is continued until convergence is achieved on the parameter estimates.

Algorithm Summary

Figure 1 is a flow chart of the described algorithm for estimation of system states and uncertain model parameters from noisy measurements. The major steps in the process are as follows:

1. Using the current best estimates of the parameters, solve the two point boundary value problem (Eq. 11) using the modal method (long data spans) or Smoothing-Riccati method (short data spans).

2. With the smoothed state estimate and Lagrange multiplier sequences from the solution to the TPBVP, calculate the gradient of the performance index with respect to the uncertain parameters.

3. If there is no a priori estimate of the inverse curvature matrix, use the rank one correction procedure with orthogonal initialization to obtain an initial approximation to the inverse curvature matrix. If there are p parameters to estimate, p+1 iterations are required.

4. After initialization of the inverse curvature matrix, use the rank one correction procedure to update the inverse curvature matrix approximation on succeeding iterations.

5. Use the parameter gradient and the inverse curvature matrix to improve the parameter estimates with a modified Newton's method procedure.

6. Repeat steps 1, 2, 4, and 5 until convergence is achieved.

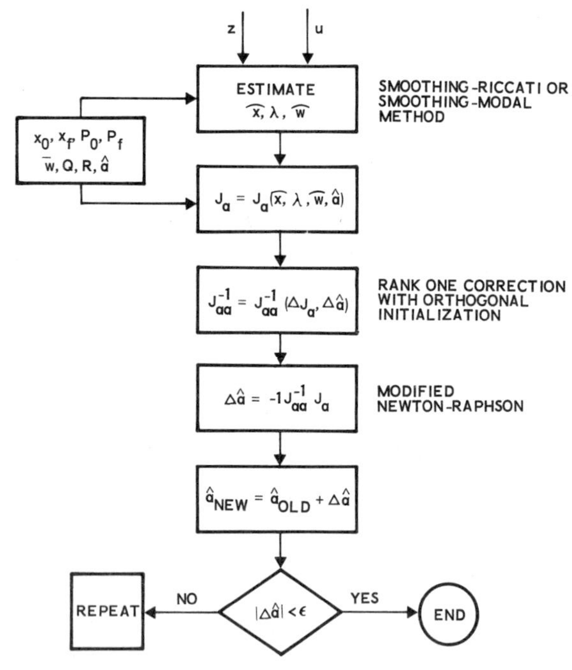

Fig. 1. Flow chart of smoothing-identification algorithm.

COMPARISON WITH THE FILTER-SENSITIVITY FUNCTION METHOD

Many parameter identification algorithms are based on forward Kalman filtering with auxiliary computation of a large sensitivity function matrix. The gradient of the performance index and an approximation of the curvature matrix are obtained from the sensitivity functions, which are partial derivatives of the estimated states with respect to the parameters (Astrom and Bohlin, 1965; Mishne, 1978). This filtering method has several advantages. First, the sensitivity functions give an approximation to the curvature matrix on each iteration. No initialization process is necessary. In addition, the filtering method can be used in situations which require online (real time) parameter identification (Mishne, 1978; Duval, 1976). However, for comparison purposes here, an offline version of the filtering method was used in an iterative manner. Sensitivity function reductions (Gupta and Mehra, 1974; Gupta, 1976) were not attempted as they are directly dependent on the system under consideration.

A side benefit from using the smoothing method for identification is that smoothed estimates of the state sequence are obtained. The smoothing method also yields an estimate of the process noise sequence, w(k), for the test run. Surprisingly, the smoothing procedure often requires fewer operations per iteration to simultaneously obtain state and parameter estimates than the corresponding filtering method.

Although the modal method of solution requires a significant amount of effort to initialize the inverse curvature matrix, the reduced cost per discrete time step in calculating the gradient of the performance index quickly overcomes any losses sustained in the initialization procedure. For example, to identify 7 parameters of a third order model with 1 measurement, 1 control input, and 1 process noise source, it would require only 20 time steps before the modal method overcame its initial disadvantage to the filtering method. Fifty steps would be required before it became less costly than the Smoothing-Riccati equation method. The number of steps necessary before cost parity is obtained varies directly as the order of the system and inversely as the number of parameters and measurements. If the order of the system were to increase to 10, the number of parameters identified to 25, but still with only 1 measurement, then approximately 40 time steps would be required to balance costs with the filtering method. An increase to 3 measurements and 2 control inputs reduces the number of steps required to 25. Corresponding values when comparing the Smoothing-Modal method to the Smoothing-Riccati method are 100 and 60, respectively.

The modal method is less expensive to use for long spans of data and the Smoothing-Riccati equation method is superior for short spans of data. Both methods provide improved state estimate sequences and estimates of the process noise affecting the system in addition to the estimates of the uncertain parameters in the system model.

AN EXAMPLE

A third order example was chosen to compare the smoothing-nonlinear programming methods with the currently popular filtering-sensitivity function method. A model of the short period longitudinal motions of an F8 aircraft (Mishne, 1978)* was discretized to obtain the following model:

$$\begin{bmatrix} q(k+1) \\ \alpha(k+1) \\ w(k+1) \end{bmatrix} = \begin{bmatrix} \phi_{11} & \phi_{12} & \phi_{13} \\ \phi_{21} & \phi_{22} & \phi_{23} \\ 0 & 0 & \phi_{33} \end{bmatrix} \begin{bmatrix} q(k) \\ \alpha(k) \\ w(k) \end{bmatrix} + \begin{bmatrix} b_1 \\ b_2 \\ b_3 \end{bmatrix} \delta_e(k) + \begin{bmatrix} \gamma_1 \\ \gamma_2 \\ \gamma_3 \end{bmatrix} \eta(k)$$

$$z(k) = q(k) + v(k)$$

Six parameters were estimated: ϕ_{11}, ϕ_{12}, b_1 and the three initial conditions. An output sample was obtained by simulation using the

*Details regarding the F8 aircraft as related to this example are described in Mini Issue on NASA's Advanced Control Law for the F8 DFWB Aircraft. *IEEE Transactions on Automatic Control*, Vol. AC-22, No. 5, October 1977.

parameter values shown in Table 1. Gauss markov purely random sequences, $\eta(k)$ and $v(k)$, were generated with a random number generation algorithm available at the Stanford Center for Information Processing. A square wave of a 1-second period and 0.05-radian amplitude was chosen as the elevator control input. The measurement data rate was taken to be 10 samples per second, with a span of 100 samples (10 seconds). This corresponds to approximately four time constants of the damped short period oscillation.

Each of the three procedures was used to estimate the state sequences, the uncertain parameters, and the initial conditions based on the same measurement sequence. Results are shown in Tables 2, 3, and 4 and Figures 2, 3, and 4. All three methods gave approximately the same answers at convergence. In this example, the

TABLE 1 Values Used in Example

$$\phi = \begin{bmatrix} 0.91503 & -0.67609 & 0.09492 \\ 0.09742 & 0.9653 & 0.005 \\ 0 & 0 & 0.95123 \end{bmatrix}$$

$$B = \begin{bmatrix} -0.85425 \\ -0.04359 \\ 0 \end{bmatrix} \quad \Gamma = \begin{bmatrix} 0.00483 \\ 0 \\ 0.0975 \end{bmatrix} \quad H^T = \begin{bmatrix} 1 \\ 0 \\ 0 \end{bmatrix}$$

$$x(0) = \begin{bmatrix} 0 \\ 0 \\ 0 \end{bmatrix} \quad x_0 = \begin{bmatrix} 0.0139 \\ 0 \\ 0 \end{bmatrix} \quad x_f = \begin{bmatrix} -0.14255 \\ 0 \\ 0 \end{bmatrix}$$

$$P_0 = P_f = \begin{bmatrix} 0.0001 & 0 & 0 \\ 0 & 1000 & 1000 \\ 0 & 0 & 0 \end{bmatrix}$$

$$a_0 = \begin{bmatrix} 0.7 \\ -0.6 \\ -0.7 \end{bmatrix} \quad A = \begin{bmatrix} 0.05 & 0 & 0 \\ 0 & 0.10 & 0 \\ 0 & 0 & 0.10 \end{bmatrix}$$

$$R = 1 \times 10^{-4}\ RAD^2/SEC^2$$

$$Q = 0.03\ RAD^2/SEC^2$$

$$\overline{\eta} = -0.035\ RAD/SEC$$

TABLE 2 Comparison of Estimated and Actual Cost per Iteration for Third Order Model

METHOD	THEORETICAL ADDITIONS	ACTUAL CPU TIME
• FILTERING	1	1
• SMOOTHING-RICCATI	0.73	0.73
• SMOOTHING-MODAL	0.58	0.66

*TIMES NORMALIZED; FILTERING METHOD EQUAL TO ONE.
TIME BASED ON IBM 370/165 DOUBLE PRECISION OPERATION TIMES.
FOR CONSISTENCY, FILTERING METHOD PERFORMED USING RANK ONE CORRECTION PROCEDURE TO OBTAIN INVERSE CURVATURE MATRIX ON ALL BUT FIRST ITERATION.

TABLE 3 Values for Estimated Parameters at Convergence

METHOD	PARAMETER ESTIMATE ERROR $\Delta\phi_{11}/\phi_{11}$	$\Delta\phi_{12}/\phi_{12}$	$\Delta b_1/b_1$	NUMBER OF ITERATIONS
• FILTERING	-0.008 (0.009)	-0.044 (0.049)	+0.008 (0.021)	8
• SMOOTHING-RICCATI	+0.001 (0.008)	+0.055 (0.068)	-0.017 (0.023)	9
• SMOOTHING-MODAL	+0.001 (0.008)	+0.065 (0.068)	-0.020 (0.023)	9

*NUMBERS IN PARENTHESES ARE ESTIMATES OF CRAMER-RAO LOWER BOUND ON PARAMETER ESTIMATE UNCERTAINTIES AND ARE OBTAINED BY TAKING SQUARE ROOT OF DIAGONAL ELEMENTS OF INVERSE CURVATURE MATRIX.

TABLE 4 Estimates for Initial Conditions at Convergence

METHOD	ESTIMATED INITIAL CONDITIONS q	α	w
• FILTERING	0.0052	0.0004	0.0278
• SMOOTHING-RICCATI	0.0043	-0.0029	-0.0149
• SMOOTHING-MODAL	0.0045	0.0079	0.0511
ACTUAL	0	0	0

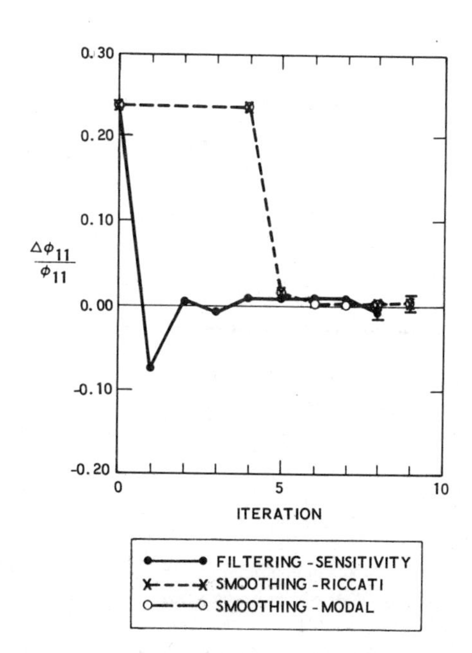

Fig. 2. Estimate of ϕ_{11} at each iteration.

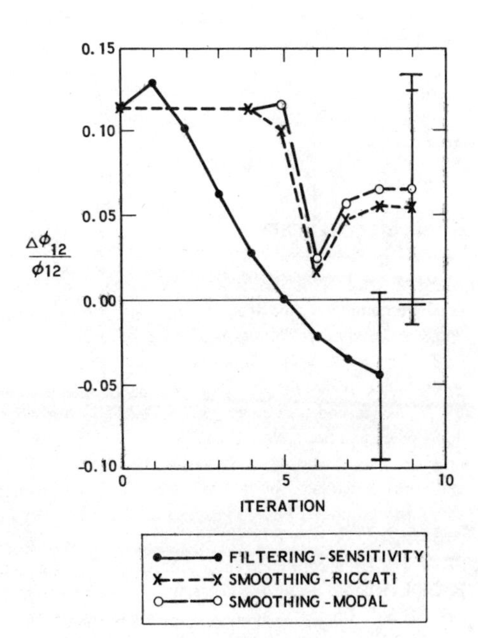

Fig. 3. Estimate of ϕ_{12} at each iteration.

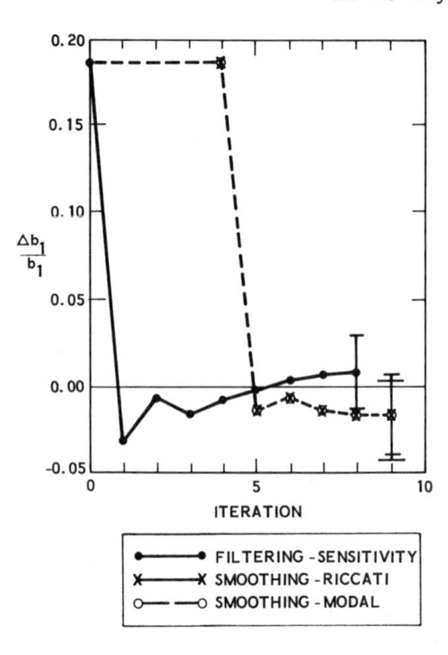

Fig. 4. Estimate of b_1 at each iteration.

filtering-sensitivity function method required one less iteration to converge than the smoothing-nonlinear programming methods. The total cost was less for the smoothing methods, however. The ϕ_{11} parameter is associated with the frequency of the short period oscillation; the ϕ_{12} parameters, with the damping factor. It is well known that the damping factor is more difficult to identify than the frequency as is evident in the uncertainties at convergence.

CONCLUSION

The nonlinear programming approach to parameter identification leads to an algorithm which provides two advantages over methods that are based on calculation of sensitivity functions for parameter identification of linear, stationary, discrete systems. First, smoothed estimates of the state vector and process noise sequences are outputs of the procedure. The smoothed state estimates are more accurate than the filtered estimates, except at the final point in the data sequence. Second, the algorithm requires fewer operations per iteration to determine the gradient of the performance index with respect to the parameters.

Use of the rank one correction procedure with orthogonal initialization overcomes the problem of estimating the inverse of the curvature matrix.

REFERENCES

Astrom, J. J., and T. Bohlin (1965). Numerical identification of linear dynamic systems from normal operating records. *Proc. 2nd IFAC Symp. on Theory of Self-Adaptive Control Systems.*

Bach, R. E., Jr. (1977) Variational algorithms for nonlinear smoothing applications. NASA TM 73.211.

Bryson Jr., A. E. and Y. C. Ho (1975). Applied optimal control. John Wiley and Sons, New York.

Chang, C. B., R. H. Whiting, L. Youens, M. Athans (1977). Application of the fixed-interval smoother to maneuvering trajectory estimation. *IEEE Trans. Auto. Control, AC-22, (5)*, 876–79.

Cox, A. B. (1979). A combined smoothing-identification algorithm. PhD Dissertation. Dept. Appl. Mech., Stanford University.

Cox, H. (1964). On the estimation of state variables and parameters for noisy dynamic systems. *IEEE Trans. Auto. Control, AC-9, (1)*, 5–12.

Duval, R. W. (1976). A rapidly converging adaptive filter for online applications. PhD Dissertation. Dept. Aero. and Astro., Stanford University.

Gupta, N. K., and R. K. Mehra (1974). Computational aspects of maximum likelihood estimation and reduction in sensitivity function calculations. *IEEE Trans. Auto. Control, AC-19, (6)*, 774–83.

Gupta, N. K. (1976). Efficient computation of gradient and hessian of likelihood function in linear dynamic systems. *IEEE Trans. Control, AC-21, (5)*, 781–83.

Luenberger, D. G. (1965). Introduction to linear and nonlinear programming. Addison-Wesley Publishing Company, Reading, Massachusetts.

Mishne, D. (1978). On-line parameter estimation using a high sensitivity estimator. PhD Dissertation. Dept. Aero. and Astro., Stanford University.

PERTUBATION-MAGNITUDE CONTROL FOR DIFFERENCE-QUOTIENT ESTIMATION OF DERIVATIVES

H. J. Kelley*, L. Lefton* and I. L. Johnson, Jr.**

**Optimization Incorporated, Jerico, N.Y. USA*
***NASA Johnson Space Center, Houston, Texas, USA*

Summary. A process for adjusting pertubation magnitude for accurate difference-quotient estimation of derivatives is described in the following. The process is intended to be carried out sequentially, alternating with iterations of a parameter-optimization algorithm. A more complex and computationally-expensive scheme for occasional auxilliary use is also described. Both adjustment schemes focus on agreement between forward and backward difference quotients. The full length paper will appear in the new journal Optimal Control Applications and Methods, Wiley-Interscience.

INTERACTIVE OPTIMIZATION SYSTEM

N. N. Moiseev and Y. G. Evtushenko

Computing Centre of the USSR Academy of Sciences, Moscow, USSR

Summary. An interactive optimization system is described which was developed for solving unconstrained minimization, nonlinear programming problems and optimal control problems. The system library of algorithms contains both novel and currently available methods. Significant extension and modifications to some of these were made in order to make them operate coherently and efficiently. A finite-dimensional representation of the control was used in order to formulate the general optimal control problems as a nonlinear program, which was solved by standard nonlinear programming algorithms. Some nonlinear programming and optimal control algorithms require the solution of additional unconstrained minimization problems. Any unconstrained minimization algorithms may be used in this case. A user can interact with the computer, monitor all system parameters, matching algorithms and their parameters to a given class of problems. The interactive capability of the computer system essentially improves, i.e. accelerates the speed of computation. The user's experience and intuition makes this process more efficient than in batch processing. A software-user interface has been created that effectively and efficiencly communicates via terminal between the user and computer in order to maximize user productivity. The interactive optimization system is used for solving various real-life control problems. Examples are given. Several optimal control algorithms were modified for solving differential game problems, and the solution of R. Isaacs brachistochrone problem is presented.

AUTHOR INDEX